The Invisible Additives

Environmental Contaminants in Our Food

Linda R. Pim

A Project of Pollution Probe

Doubleday Canada Limited,
Toronto, Ontario
Doubleday & Company, Inc.,
Garden City, New York

1981

Library of Congress Catalog Card Number 80-2430

Copyright © 1981 by Linda Rean Pim

All Rights Reserved

Printed in Canada by Webcom Ltd.

The paper in this book contains post-consumer waste.

Typesetting by ART-U Graphics

Chapter 6 first appeared in a similar form in the July/August
1980 and September/December 1980 issues of the *Probe Post*.

Canadian Cataloguing in Publication Data

Pim, Linda R., 1952-
    The invisible additives

Bibliography: p. 259
Includes index.
ISBN 0-385-17001-7 (bound). —ISBN 0-385-17002-5 (pbk.)

1. Food contamination.   2. Pesticide residues in food.
3. Food adulteration and inspection—Canada.   I. Title.

TX531.P55   363.1'92   C81-094372-7

The Invisible Additives

If you thought that food additives ended with the list of chemicals on the side of the package, think again. Many of the real and potential health hazards linked to chemicals in food stem from the contamination of our daily bread by substances in the environment. Eighty percent of all cancer is thought to be of environmental rather than hereditary origin. Our food must be examined as a possible cause of cancer and other health problems.

Until now, information on environmental contamination of food has been hiding in scientific journals and government studies, inaccessible and unintelligible to the layperson. This book brings the problem out into the open. It provides the reader with solid facts and figures about these "invisible" additives: pesticide and drug residues; poisons naturally present in some foods; pollutants such as PCBs, dioxins, lead, and mercury. It provides the consumer with tips on how to avoid certain additives and how to cut down on overall dietary intake of these chemicals. And it explores what governments should do on our behalf to ease the problems.

In the words of consumer columnist Nicole Parton, *The Invisible Additives* is "must reading for all consumers interested in the safety of our food."

Linda R. Pim is an environmental biologist and has worked on the Pollution Probe research team since 1976. Her research into the problem of food-processing additives led to her first book, *Additive Alert: A Guide to Food Additives for the Canadian Consumer,* published in 1979. Since then she has spoken to many consumer and professional groups on the subject of chemicals in food. She is currently working on a revised edition of *Additive Alert.* Linda lives near Toronto, Ontario. She is an avid vegetarian cook and organic gardener.

Other Books from Pollution Probe/Energy Probe

For my sisters
Barbara, Elizabeth, and Carolyn

Foreword

MOST JURISDICTIONS in Canada rank toxic chemicals among our top environmental concerns. There is a myriad of ways in which toxic and potentially toxic chemicals can affect living systems, including humankind, ranging from biological magnification in ecosystems, through the largely unknown effects of exposure over many years to trace amounts of toxic chemicals, to acute lethal toxicity. One of the most important routes by which chemicals enter living systems is the food that all forms of life must consume to provide energy for growth, for reproduction, and for all the other processes that characterize life itself.

There are two basic ways by which chemicals that are not inherently a natural part of the food itself can enter the materials that we eat. The first is by their deliberate addition to food substances to achieve some supposedly desirable objective, such as preservation, improved appearance or texture, and enhanced nutritive value. Although we know distressingly little about what effects these additives will have on us in the long run, at least we know that they are in our food because they were put there for a purpose. In an earlier book, *Additive Alert,* Linda Pim gave a penetrating discussion of this particular issue.

The second way by which chemicals enter our food is much more insidious than deliberate addition. It is by accident. Many chemicals the use of which is completely divorced from human nutrition have the capacity to move through the environment and to enter food chains, including the one on which we depend. Pesticides represent a special category of this class of chemicals: their use is deliberate and closely related to food production but the fact that some of them enter our food after they have done their job of protecting it from insect pests and plant diseases is unintentional—an accident.

Personally, I find the presence of this second group of chemicals, *The Invisible Additives,* even more disturbing than the deliberate additives. At least with the latter if we discover that one of them has some dangerous or unpleasant effect we can control the situation simply by stopping its addition to our food.

With the second group we are dealing with chemical wastes, the flotsam and jetsam from the sea of chemicals we use for an array of purposes, from lubrication and insulation to making paper and pest control. Often we do not even know the specific source of these hidden additives in our food, and therefore lack even the most primitive means to control them except by the bludgeon of outright bans on their use for any purpose whatever.

In this her second book, *The Invisible Additives,* Ms. Pim has maintained the high standard she established with *Additive Alert.* Her research is thorough, and her approach balanced and objective, though she makes no bones about her bias towards safety and caution. And she strongly believes that the onus for the burden of proving safety should be with those who manufacture and use chemicals destined to become invisible additives to our food.

The Invisible Additives is a thoroughly professional book which should be read by everyone concerned about the chemical contamination of our food. And who can afford not to be concerned?

D. A. Chant
Director, Joint Study Centre for Toxicology
University of Toronto

Contents

Acknowledgements

IN RESEARCHING AND writing this book, the people I have talked with have been as important as what I have read. (Three cheers, nevertheless, for the University of Toronto's Science and Medicine Library.) I have felt like a vehicle being fuelled by the knowledge and opinions of almost innumerable individuals. Among these, I mention the following and express my thanks to them:

Pollution Probe staff researchers Moni Campbell, Bill Glenn, and Anne Wordsworth, and Energy Probe researcher Norman Rubin;

Pollution Probe volunteer researchers Gina Bem (toxins in natural foods—Chapter 3), Dominic Romano (international regulation of food contaminants—Chapter 4, and pesticides in imported produce—Chapter 5), Carol Brotman (heavy metals—Chapter 10), and Marilyn Tate (occupational exposure to pesticides—Appendix A);

Numerous officials in the Health Protection Branch of Health and Welfare Canada, including, in Ottawa: Dr. Earl Coffin, Dr. Harry McLeod, Dr. R.R. MacKay, Dr. T. Kuiper-Goodman, Ross Elliot, Dr. Robert Ripley, Dr. Peter Bennett, Dr. Ian Munro, Diane Kirkpatrick, Dr. B.H. Lauer, and Dr. Peter Scott; in Toronto: Stephanie Charron, Marilyn Knox, and Pat Robinson;

In Agriculture Canada, Dr. Christopher Young (Chemistry and Biology Research Institute, Research Branch), Dr. E.A.G. Hagley

(Research Station at Vineland, Ontario), and Wayne Ormrod (Plant Products Division, Food Production and Marketing Branch);

In the Ontario Ministry of Agriculture and Food, Dr. Richard Frank (Director, Provincial Pesticide Residue Testing Laboratory), Dr. A.A. van Dreumel (Veterinary Services Branch), and Glen Ward (Milk Industry Section, Farm Products Inspection Branch);

Doug Wilson, Dr. D. Harding, and other members of the Pesticides Advisory Committee (Government of Ontario);

Gordon Van Fleet (Pollution Control Branch, Ontario Ministry of the Environment);

Peter Crabtree (Waste Managment Advisory Board, Government of Ontario);

Chris Taylor and Monte Hummel (Environmental Studies, Innis College, University of Toronto);

Dr. Helen Liu (Department of Environmental Biology, University of Guelph);

Dr. J.G. Stopps (Faculty of Medicine, University of Toronto);

Dr. Tom Hutchinson (Department of Botany and Institute for Environmental Studies, University of Toronto) and his metals laboratory;

Dr. Donald Chant (Director, Joint Study Centre for Toxicology, University of Toronto/University of Guelph, and former Chairman, Board of Directors, Pollution Probe Foundation);

Dr. Ross Hume Hall (Department of Biochemistry, McMaster University, Hamilton, Ontario) and Beatrice Trum Hunter (Hillsboro, New Hampshire);

Dr. Stephen Safe (Department of Chemistry, University of Guelph, Guelph, Ontario);

Dr. Stuart Hill and Toni Bird at the Ecological Agriculture Project, Macdonald College of McGill University, Ste Anne de Bellevue, Quebec;

The folks at CBC "Radio Noon" (Toronto)—Bruce Rogers, Carol Giangrande, Roy Maxwell, Barbara Klich and crew—who provide an interesting and informative news service on agriculture, food, and health issues;

Neil Robson (Toronto, Ontario), Peter von Stackelberg (Regina, Saskatchewan), Michael Levenston (Vancouver, British Columbia), Adele Hurley (Toronto, Ontario), Canadian Organic Growers (Toronto, Ontario), David Skinner (Ganges, British Columbia), and Jack Warnock (Naramata, British Columbia).

The following people read all or parts of the manuscript and offered their comments and suggestions:

Moni Campbell, Bill Glenn, Anne Wordsworth, Judy Liefschultz, Rick Beharriell, and Dr. Donald Chant, all of Pollution Probe;

Larry Solomon, Norman Rubin and Dr. David Brooks, all of Energy Probe;

Dr. Richard Frank, Dr. Stephen Safe, Dr. Ross Hume Hall, Carol Brotman, Dr. Stuart Hill, Dr. Christopher Young, Neil Robson, Adele Hurley, and David Skinner;

Nan Martin (Oxford, England) and Carolyn Pim (Calgary, Alberta).

Thanks to the people without whom the manuscript could not have become a book: Grace Deutsch who did the copy-editing; Bob Garbutt, the book's designer; and Roxshana and Rumi Khambata who did the typesetting.

Special thanks to Jim, Kathy, Steve, Nora and Gord, and Carol for their support, encouragement, love and/or other appropriate sentiments; and to "Slug," my low-tech, energy-conserving typewriter who misspells when I least expect it.

Finally, very kind words are in order for Rick Archbold (editor) and Carolynne Hastings (publicity manager) at Doubleday Canada, whose hard work is exceeded only by their good cheer.

LINDA REAN PIM
Thornhill, Ontario
January 1981

About Pollution Probe

Pollution Probe, founded in 1969, is an independent registered charitable foundation and one of Canada's leading public interest groups. Probe has been at the forefront of many of Canada's central environmental issues for over a decade. The organization has been responsible for significant steps forward in alleviating air and water pollution, curbing the generation of solid waste, promoting recycling, adopting better land use policies, and implementing stricter control on toxic chemicals in our environment.

Projects of the Foundation are the Pollution Probe research team, Ecology House (12 Madison Avenue, Toronto, Ontario, an urban demonstration of conservation in the home) and the *Probe Post* (a bimonthly newspaper).

This book was written as part of Pollution Probe's ongoing work in the area of human exposure to environmental contaminants.

For more information on Probe, write to:

Pollution Probe
University of Toronto
Toronto, Ontario
M5S 1A1

The
Invisible
Additives

PART ONE

An Overview

1 Food Additives Revisited

IN AN ERA when sweatsuit-clad business executives puff through their lunch hours on the jogging track, commuters exchange their transit tickets for ten-speed bicycles, and bran muffins encroach on the domain of the doughnut as a quick snack, public interest in the safety of food additives is not surprising. All these shifts in lifestyle indicate a new appreciation of the responsibility each of us must assume for our own good health. What it really boils down to is preventive medicine with a fresh face.

And there is a great deal of preventing to do in our modern industrial society. Prosperity and technological progress have brought with them some unfortunate side effects. While the food-processing additives (colours, preservatives and so on) that have been in the limelight recently are a significant threat to our health, this book is concerned with a more serious and insidious source of chemicals in the food we eat: environmental additives.

These environmental additives are chemicals which inadvertently contaminate our food in the course of its passage from farm to dinner plate. Their appearance in food is incidental, inconsistent, and often difficult to monitor. The names of these chemicals won't be found on the side of a food package. In a sense, they are *invisible* additives.

It is the food-processing additives—the visible additives on food labels—which have stolen the food risk show. Many consumers have been looking at the processing chemicals as the end of the additive discussion. We now find a burgeoning squadron of additive-conscious consumers avoiding, or at least raising doubts about, the chemical cornucopia offered as standard fare in the centre aisles of the super-market. But these processing chemicals are really only the beginning of the additive story. Consumers generally have a much shakier appreci-ation of environmental additives, since food labels remain mute when it comes to chlordane (a pesticide residue in milk), aflatoxin (a poisonous mould on peanuts), and sulfonamide (a drug residue in pork).

Efforts by government agencies to educate the public about food additives are primarily, if not exclusively, directed to processing chemi-cals. For example, Health and Welfare Canada's food additive educa-tion program deals entirely with processing additives.

But the layperson often does not make a clear distinction between processing and environmental food additives. In the summer of 1979, the federal health department canvassed 25,000 Canadians for their opinions on food additives and found that a whopping 52 percent included pesticides in food as a type of food additive. The government does not; it has no definition of environmental additives; it defines only intentional or processing additives. (The legal definition of the latter is shown in the box.) Furthermore, when those canvassed were asked what their greatest concern was about the safety of the food supply, 46 percent named pesticides and industrial pollutants, while only 35 percent named processing additives.

If the federal government won't publish educational material on environmental additives, someone must. Enter Pollution Probe. *The Invisible Additives* is the necessary companion volume to *Additive Alert,* Probe's handbook on food-processing chemicals.

This book may raise as many questions as it attempts to answer. It is almost impossible to arrive at definite conclusions on this subject when the impacts of environmental food contaminants are still so poorly understood. We're a long way from having the last word on most of these chemicals. DDT, that kingpin of insecticides, has been in use for 35 years. Like most carcinogens, its health effects may take decades to show up. What will happen after, say, 36 years of exposure?

With this information bank desperately in need of more deposits, controversy about the extent of the health risks of small amounts of contaminants in our food rages unabated among scientists, regulatory agencies, the agriculture and food industries (collectively, "the experts"), and lay consumers. Making a decision to allow or restrict the use of a purportedly toxic chemical is agonizingly difficult when the benefits and risks of using (or not using) it are so uncertain.

Legal Definition of Food Additive (Canada)

"'food additive' means any substance, including any source of radiation, the use of which results, or may reasonably be expected to result in it or its by-products becoming a part of or affecting the characteristics of a food, but does not include

(a) any nutritive material that is used, recognized, or commonly sold as an article or ingredient of food,
(b) vitamins, mineral nutrients and amino acids,
(c) spices, seasonings, flavouring preparations, essential oils, oleoresins and natural extractives,
(d) agricultural chemicals,
(e) food packaging materials and components thereof, and
(f) drugs recommended for administration to animals that may be consumed as food."

Source: Regulation B.01.001, *Food and Drugs Regulations.*

The more complex and technical the food contamination debate becomes, the more alluring to the experts is the belief that the average citizen cannot be trusted to understand the information that *is* available. But surely it is the experts' responsibility to provide that information to laypeople and help them interpret it? In turn, surely it is the responsibility of laypeople to become well enough informed to be able to guide regulatory bodies to the most socially acceptable decision, one based on the informed consent of the public rather than on its ignorance?

A recent editorial in *Not Man Apart*, the journal of the United States arm of Friends of the Earth (an international public interest environmental organization), addressed the issue of the lay public's need to become informed. Editor Harry Dennis wrote as follows: "People generally react in one of a few different ways to news of toxic chemical problems. Most prefer not to talk about it. Perhaps they don't want to take it seriously because they don't believe the problems are all that bad; maybe they think they are so bad that nothing can be done about them. But sticking your head in the sand doesn't help. Becoming informed does."

The Invisible Additives is meant to inform the lay public about environmental contaminants in food and to interpret the controversies surrounding these chemicals. The remainder of Part I (Chapters 2 through 4) provides an overview of the problem of food contamination and its regulation by government. Part II (Chapters 5 through 11) looks at the key groups of environmental additives one by one— pesticides, animal drugs, poisonous moulds, organic pollutants, metal pollutants, and a miscellaneous category. Each of these chapters ends with a section suggesting what you can do to curb the intrusion of each group of environmental toxins into your daily bread.

Part III (Chapter 12) sums up what we've found out in the course of the book. The appendices include a comprehensive glossary (Appendix B), a chapter-by-chapter list of sources (Appendix C), and a general guide to obtaining further information (Appendix D).

This is a *hand*book. Read it . . . then use it!

2 Chasing Zero Risk

AS LONG AS you get out of bed in the morning, life is fraught with risks. (There may even be some if you don't.) Some hazardous situations in which we find ourselves, such as a pack-a-day cigarette habit or once-a-week purchase of lottery tickets, are entirely of our own doing. Other risks are taken less willingly—most of us are forced to breathe pollution-clogged city air, for example. We are subjected to still other risks quite unknowingly. People slipped saccharin into their coffee for 80 years before laboratory evidence indicated any adverse health effects.

Recent psychological research has shown that we often perceive the risks in our lives in ways which distort the actual dangers involved. For example:

● Appreciation of risk is greatly influenced by the level of publicity given to particular events and by the number of people affected by any one event. A jumbo jet crashes and 300 people lose their lives—a tragedy, no doubt. But it is no more tragic than the death of the same number of people, one by one, in hospital wards across the country.

● We are a thousand times more willing to accept voluntary risks than those to which we are involuntarily exposed. While some of us are content to assume the substantial risks of jaywalking through rush-hour traffic, we find the lung-damaging vapours from automobile exhausts to be highly unacceptable.

• The sooner that a risk taken has a harmful effect, the keener is our perception that taking the risk is, indeed, a gamble. For the unlucky jaywalker, the risk is well recognized and the result instantaneous. For the consumer of trace quantities of a cancer-causing pesticide residue in food, the risk of developing cancer 30 years down the road may seem nebulous.

Risks of the Chemical Environment

The above perceptions colour our attitudes and influence the actions our society takes to restrict the presence of harmful chemicals and other risks in our environment. Before going any further, though, let's clarify the meaning of the three buzz-words most used in discussions of environmental chemicals: toxicity, hazard (or risk), and safety.

The *toxicity* of a substance is simply its capacity to produce biological damage of one kind or another, from allergies to cancer. The *hazard*, or *risk*, associated with that substance is the probability that injury will result from its use in a proposed amount and manner. So, for example, water (a chemical, like everything else around us) would be toxic if you were to drink up Lake Superior at one sitting, but it does not present a hazard because in the small quantities in which we normally drink it, water performs an indispensable function in our bodies. When there is no hazard, we have *safety*, or the practical certainty that injury will not result from the proposed use of the chemical.

To a great extent, advances in technology, such as the invention of industrial pollution control equipment, have helped us prevent highly toxic amounts of chemicals from entering the air we breathe, the water we drink, and the food we eat. It is unlikely, for example, that there will be a repeat performance of the deadly London smog of 1952. Cause and effect were clearly linked. In the 1980s, the focus in the chase towards zero environmental risk is on the health effects of the seemingly small, or "trace," quantities of toxic substances that enter our bodies every day without immediate, obvious harm.

In the early decades of this century, it was discovered that trace amounts of the chemicals now called vitamins were essential in the human diet in order to *prevent* certain diseases. It didn't take much vitamin C to prevent scurvy or vitamin D to stave off rickets. Today, we are turning the tables and asking whether traces of toxic chemicals can *cause* disease. In a few instances, there is conclusive evidence that this can happen. For example, diethylstilbestrol (DES), a growth-promoting hormone at one time fed to beef cattle, and aflatoxin, a mould contaminant that shows up in nuts and grains, are both capable of inducing cancer in laboratory animals when present as *one-billionth* of the diet (that is, a concentration of one part per billion), a level sometimes found in the human food supply.

In more cases than not, the technology for detecting trace amounts of toxins entering the human body has far outstripped our understanding of what health impacts these traces might have. For example, analytical equipment can measure infinitesimal concentrations of pesticide residues in food, yet we remain uncertain of the effects of eating those minute amounts three times a day over the course of a lifetime.

Tests on well-nourished and otherwise pampered laboratory animals who are fed mega-doses of individual toxic chemicals provide only a limited amount of information about how these substances will act at lower doses in a less controlled environment. The tests are, however, usually predictive. For example, all of the 26 chemicals directly proven to cause cancer in humans are also carcinogenic in experimental animals, so we could expect that the reverse might hold. But many scientists believe that laboratory studies very likely understate the risks to humans, allowing a number of "false negatives" to slip by unnoticed. A chemical is a false negative when it appears non-toxic under test conditions but is toxic once it leaves the laboratory and exerts its effects on a human being experiencing several other environmental stresses at the same time.

There is, then, a credibility gap between the laboratory science of "classical" toxicology and the real-life exposure of people to chemicals in the environment. What is called for now is emphasis on "ecotoxicology"—the scientific consideration of the health impacts of several chemicals at once on people eating less than ideal diets amid the pressures of modern living. Under such conditions, traces of toxic chemicals thought to be of minimal significance to the health of a laboratory rat can take on real importance in the human environment.

Evidence of the ecotoxic effects of chemicals is scanty because ecotoxicology is a new and underdeveloped field of study. But it has been discovered, for instance, that many pesticides increase sharply in toxicity when animals are fed protein-deficient diets. Therefore, pesticide exposures considered safe in protein-glutted industrialized countries may very well be unsafe in less developed nations where dietary protein deficiency is common.

Of Quantities and Culprits

Trying to quantify the health risks of chemical traces in the environment is difficult, if not impossible. The probability of being hit by a meteor is very low, but at least it is known. The probability of contracting disease from small doses of a certain food additive or air pollutant *may* also be low, but it is unknown. There may be little point in trying to pin down the chance at all, since the figure would be rendered meaningless in the face of interactions among the chemicals around us, interactions that may increase or decrease the effect of each one. We must consider synergy, the unpredictable situation in which the toxicity of two substances occurring together is greater than the sum of their separate effects.

Environmental risk assessment is further muddied by the fact that effects may take years or decades to show up. The International Agency for Research on Cancer believes that about 80 percent of human cancer has environmental rather than hereditary causes. Cancers may take 20 or 30 years to develop. Efforts to isolate the cause of a particular malignancy in 1980 from someone's chemical environment in 1950 are painstaking at best. As American public interest food activist Michael Jacobson wrote recently in *Science*: "The exact number of deaths due to cancer-causing additives cannot be calculated with great accuracy, because, thoughtless and inconsiderate to policy-makers though it may be, tumors do not bear tags identifying their cause or causes."

The inescapable conclusion, then, is that it makes little sense to require absolute proof of harm from traces of toxic chemicals before taking practicable measures to reduce the risks. We could wait until 2001 for indisputable evidence. All that should be necessary for action is some reasonable indication from scientific research that a risk is likely to exist when the chemical is used.

Weighing Benefits Against Risks:
What's Acceptable . . . and to Whom?

Zero risk from chemicals in our environment is probably unattainable, but that's a poor excuse for not trying to chase it. Of course, it would be much easier to eliminate chemical hazards if we weren't constantly required to juggle the risks with the benefits these chemicals confer, or to provide safer alternatives to toxic chemicals now judged to be serving a necessary function in our society.

In many facets of our daily lives, we undergo a risk in order to achieve a benefit which we perceive to be greater than the risk. If you must travel from Vancouver to Winnipeg quickly, you accept the risk that the jet might crash on landing. The risk is deemed acceptable to most, since there is simply no alternative means of achieving the benefit of a fast flight.

On the other hand, many people are eating less bacon and ham than they used to, wary of the risks of sodium nitrite, the ubiquitous colouring and preservative for processed meats. For some, nitrite's benefits—meat safe from botulism—and risks—carcinogenic nitrosamines formed from the nitrite—are seen to be on a more or less equal footing. Therefore, with other, safer, protein sources available, the risks of such meat are concluded to be unacceptable.

It is as difficult to quantify benefits as it is risks. As Health and Welfare Canada official Harold Grice has put it: "Economic gains expressed in dollars, convenience expressed in unknown units, aesthetic appeal in numbers of any kind and palatability expressed in yet-to-be agreed upon measure, are examples of measurement units that make objective appraisal of benefits all but impossible. Because of this, regulatory agencies which are required to make legislative decisions rely to a large extent on subjective impressions in arriving at benefit decisions."

Even if the probabilities of environmental risks were known, we would still be left with the question: which is of greater benefit to society—not using the chemical and thereby avoiding harm to human health, or using the chemical and reaping its benefits (insects exterminated, meat preserved)? When we do try to quantify the problem, the results can be ambiguous and not all that helpful to decision making, as is shown by the example of the mythical Veg-E-Wax.

"VEG-E-WAX"
A Cost-Benefit Analysis

"Consider a fictitious new product, Veg-E-Wax, designed to coat fresh fruits and vegetables. Its demonstrated advantages are reducing losses in storage and preserving nutritive value. Aside from the cost of application, its disadvantages are making food look less appetizing and possibly causing cancer to workers who apply it and to consumers who fail to wash fruit. A highly simplified cost-benefit analysis of the decision to apply Veg-E-Wax to a $10 million (market value) shipment of pears bound for storage might appear as follows:

	$ million
Advantages (benefits)	
Guaranteed reduction in storage loss from 30% to 20%	1.0
Improved nutritive value (translating into a 10% increase in market value in the 80% that is not lost in storage)	.8
Total benefits	1.8
Disadvantages (costs)	
Cost of application	.1
Cancer in .1% of 100 workers (at $1 million per case)	.1
Cancer to users (1 million consumers, of whom 10% fail to wash fruit, of whom .0001 contract cancer as a result, at $1 million per case)	.1
Unappetizing appearance (20% loss in market value of pears not lost in storage)	1.6
Total costs	1.9

"In this calculation, the costs slightly outweigh the benefits and the packer should decide not to use Veg-E-Wax. The viability of this conclusion depends upon its capacity to withstand small changes in the figures. If there were only an 18 percent loss in market value due to waxy look of the fruit (translating into a cost of $1.44 million), the balance would tip the other way. It might be impossible to predict this loss with the precision needed to take confident action.

"Even larger effects may accompany changes in fundamental assumptions. A packer with no social conscience might decide not to worry about the $200 000 in cancer costs, reducing total costs to $1.7 million. Other interested parties, such as consumers interested in maximizing value and minimizing personal risk, might structure the problem entirely differently."

Source: B. Fischoff, P. Slovic, and S. Lichtenstein, "Weighing the Risks: Which Risks Are Acceptable?" *Environment,* May, 1979, pp. 17-20, 32-38.

Now we can begin to see why our newspapers are peppered with conflicting reports about the risks and benefits of chemicals in the environment. We are hampered by an inadequate base of scientific information, yet there is tremendous political pressure on scientists to come up with clear-cut interpretations of inconclusive scientific evidence. The reports are at odds with each other less because the information on which they are based conflicts than because the scientists, regulatory officials, industry representatives, and consumer advocates all tend to fall back on their own judgments of benefit, risk, and safety when the hard data run out. Pronouncements cloaked as scientific truths may be more laden with the values of the individual than he/she would have us believe.

Those values, in turn, are coloured by one's position in the political structure of society. Who reaps the benefits and who gets stuck with the risks? Jerry Ravetz, a scientist who writes about the history and philosophy of science, has developed the concept of the "risk triangle." (It is equally valid as a "benefit triangle.") He notes that there are three sides involved in every hazard: those who create it, those who experience it, and those who regulate it. All three sides of the triangle may come together in the same person (such as a mountain climber). But for most risks in industrialized society, the sides are separate. So, for example, a food chemical company produces saccharin (creates the risks of marketing a carcinogen); a diabetic or dieter buys and uses it (experiences the risk); and the Canadian federal government bans its use in foods, restricting its availability to tabletop sweeteners (regulates the risk).

Everyone eats, so all sides of the triangle experience the risks of toxic contaminants in food. But those who receive extra benefits from the presence of chemical traces are likely to downplay their risks, while those with intimate technical knowledge of the risks may seem, to the first interest group, to be overstating the dangers.

An *Atlantic Monthly* report on chemicals used in meat production put it this way: "The definition of safety often takes on political implications as the food lobbies maneuver to protect their investment. The decisions are therefore as politically hazardous for the regulators as they are potentially perilous for the public. To the sausage makers and the cattlemen who use the chemicals to make their hot dogs red or their steers grow faster, the risks may seem very remote and the benefits very concrete; but to the cancer researcher the same benefits may seem trivial and the risks—a possible increase in the incidence of cancer a decade or more after human exposure—catastrophic."

Reducing Chemical Risk—How Much?

The economic benefits to society of many chemicals are clear and the health risks are often ill-defined, so the incentive to gamble with the safety of a food additive is great. But as more and more evidence of the health impacts of chemical technology tumbles in, many traditional values are changing. There is increasing effort to demonstrate safer means of achieving the same end, the same benefit with less risk. Ecosystem management of agricultural insect pests (Integrated Pest Management) is a practical way of achieving the same result as widespread application of toxic chemical pesticides—pest-free crops—with fewer chemical residues.

On the other hand, since it is apparent that not all the hazardous chemicals around us have ready-made alternatives waiting in the wings, we must reconsider how badly we need or want those chemicals' functions. When certain functions cannot be performed safely enough to be acceptable to most of society, then society may decide to eliminate the function entirely. Some benefits we can all live without. For example, how important is it that the skins of oranges be free of surface blemishes? We spray oranges with pesticides to rid them of damage by thrips, insects which disfigure the peel but have no effect on the part of the orange that we eat.

How important is "cheap food"? Is it really all that cheap? In defence of the status quo, federal Agriculture and Health and Welfare officials often proudly trot out figures indicating how small a percentage of our incomes we Canadians spend on food. Food prices are a volatile issue; anyone suggesting that costs need to be increased, for whatever reason, is soon shown the door.

But, as the Science Council of Canada put it in a 1979 report on the Canadian food system, "cheap food appears to have hidden costs. These are manifest as poor returns to many farmers and fishermen, health problems in the population, and degradation of the productive resource base [soil erosion, depletion of fish stocks, etc]." In other words, if we add all the extra social and environmental costs of our current diet to the actual pricetag shown, our food is not all that cheap. Introducing more environmentally benign ways of producing, processing, and distributing food is advocated by environmentalists precisely because its total cost to society is thought to be "cheaper."

Psychologist Baruch Fischhoff and his associates at Decision Research Incorporated in Oregon have looked deeply into better ways of managing risks in our environment. There is much room for improvement in the scientific understanding of chemical risks, the education of the public on these risks, and the streamlining of regulatory institutions to better handle them. Fischhoff concludes that "we may have to decide to live with some hazards, reduce others, and do without some technologies which entail hazards; and these decisions will have to be made in the context of substantial uncertainties."

The uncertainties surrounding the subject of environmental contaminants in our food will be evident in the pages that follow. But there should be no question of the value I place on the protection of human health in the process of supplying our daily bread. With other sectors of society, such as the chemical and food industries, trumpeting the benefits of food chemicals through high-budget public relations programs, it is imperative that environmental and medical researchers balance those claims with information on the health risks of these same chemicals. That is the central purpose of this book.

3 A Catalogue of Food Risks

"THE CONTENTION THAT the food supply is killing us is not supportable." So says Dr. Alex Morrison, head of Health and Welfare Canada's Health Protection Branch. That judgment is debatable. The fact is that there *are* environmental risks in eating for a living.

The entries in this catalogue are *not* ordered according to degree of risk. The first four will not be covered elsewhere in the book, consisting as they do of risks other than contamination by environmental chemicals, but deserving mention nevertheless. The remainder of the catalogue is an outline of things to come.

Malnutrition

In Canada? Yes! In a country as developed as Canada, it is unsettling to have to accept the fact that malnutrition is widespread. Unlike the malnutrition we associate with food shortages in Third World nations, poor nourishment here results from eating the wrong kinds of food. Generally, we are overfed but undernourished!

In 1972, the federal government's Nutrition Canada survey of food habits and health status found that half of adult Canadians are overweight (as much due to a sedentary lifestyle as to excessive caloric intake) and that dietary deficiencies of iron, calcium, vitamin D and some of the B vitamins are common.

Further hazards of the highly refined and processed North American diet were summarized in 1977 by U.S. Senator George McGovern's Select Committee on Nutrition and Human Needs. The food we eat every day, the committee said in its controversial report, "may be debilitating or deadly. The amounts of fats, saturated fat, cholesterol, sugar and salt being consumed may be major factors in the incidence of heart disease, bowel, breast and colon cancer, diabetes and stroke, to name some of the most widespread diseases that have been related to diet. These dietary elements, like other substances having an adverse cumulative effect on the body, appear innocuous, even appealing in individual portions. However, their impact may be devastating."

In 1980, the U.S. Departments of Agriculture and of Health, Education and Welfare took McGovern's findings to heart and issued dietary guidelines for American citizens. The advice, which applies equally to Canadians, includes increasing consumption of fruits, vegetables, and whole grains to obtain required vitamins, minerals, and fibre; reducing the intake of fat- and cholesterol-rich foods to stave off heart disease; and avoiding excessive sweetening and salting of food, which affect dental health and high blood presure, respectively. In short, a tall order—and as critical to our health as the concern for reducing our intake of chemical additives.

Processing Additives

Included here are the colours and flavours that make Tang what it is today (a lot of sugar for the most part), the stabilizers that prevent ice crystals forming on your brick of buttered pecan, and the preservatives that hold bread, crackers, cookies, and oils in suspended animation for months.

While a good majority of these additives appear to have no harmful health effects, a number of them *are* of dubious safety. Moreover, their use in food is often only for cosmetic purposes and serves no utilitarian function. The reader is referred to *Additive Alert* for a more comprehensive look at processing chemicals.

Microbiological Contaminants

A windy way of saying food poisoning. Each year, some 400,000

Canadians contract food poisoning, courtesy of a number of species of unsavoury bacteria. Most common are those of the *Staphylococcus* and *Salmonella* groups; rare but deadly is the bacterium *Clostridium botulinum*, the cause of botulism.

High-risk foods consist of animal products that are improperly prepared and stored—not enough cooking or not enough refrigeration. They include poultry, meat, processed and canned meat, milk and milk products, and shellfish. Beware the bulging tin can, too—a portent of botulism. Also, don't feed honey to a baby, since the honey may harbour spores of *C. botulinum* as well. (*Clostridium* in honey does not affect adults.)

Health and Welfare Canada has sound advice on the microbiological front: "When in doubt, throw it out." Check the source list for this chapter (in Appendix C) for further tips.

Natural Toxins

Apologists for highly fabricated, additive-laden food products often use the "natural toxin ploy" to divert the public's attention from harmful processing additives. "There's cyanide in apple seeds, you know." They're perfectly correct, of course. A number of common unprocessed foods have poisonous substances in them, not as contaminants, but as part of their tissues, their biochemical makeup (see *Table 1*).

In most cases, levels of the toxins are very low in relation to an individual's normal consumption of the food. In other words, unless you're in the habit of sprinkling a cup of apple seeds on your morning granola, cyanide is not a major worry, at least in terms of immediate health effects. It is unclear, however, what the long-term impact might be of taking in small amounts of these naturally occurring toxicants over the course of a lifetime.

In some cases, *no* amount of the food should be eaten. For example, the oxalate in rhubarb leaves prevents their addition to a tossed green salad, and some beans (soy beans, lima beans, chick peas) should never be eaten raw.

Pesticides

Chemicals used to control pests (crop-eating insects, weeds, and other undesirables) on the farm are applied according to schedules designed to prevent significant chemical residues in food once it reaches the dinner table. But it happens that pesticides may be applied at too high a rate, or too close to harvest time, or both, with unacceptable residues remaining in food. Some pesticides degrade so slowly that they may appear in food years after they were last applied in the fields. (See Chapters 5 and 6.)

Table 1 / *Natural Toxins*

Food	Toxin and its effects
Grains	
millet	hydrogen cyanide (a respiratory inhibitor)
sorghum	hydrogen cyanide nitrosamines (carcinogenic)
wheat (flour)	trypsin inhibitor (trypsin is an enzyme important in the digestion of protein)
Vegetables	
beets	high in nitrate (causes methemoglobinemia, a blood disease, and can convert to nitrite and then to nitrosamines)
cabbage	contains a goitrogen (causes goiter)
carrots	contain a goitrogen
chick peas (raw)	contain a goitrogen trypsin inhibitor lathyrism (a disease of the nervous system)
lettuce	high in nitrate
lima beans (raw)	hydrogen cyanide trypsin inhibitor
mushrooms (some species)	toxic to liver, kidneys, heart, muscles, nervous system
peas	hydrogen cyanide
potatoes	flesh—trypsin inhibitor green skin and sprouts (due to light exposure)— solanine (an inhibitor of cholinesterase, an enzyme of the nervous system) stems and leaves—alkaloids (cause severe digestive disorder)
radishes	high in nitrate
red peppers	amide capsaicin—irritant

Food	Toxin and its effects
rutabagas/turnips	contain a goitrogen
spinach	contains a goitrogen high in nitrate
sweet potatoes/yams	contain a goitrogen trypsin inhibitor hydrogen cyanide alkaloid convulsant (*Discorea hispida*)
tomatoes	juice—lycopene (accumulates in liver; toxic in large quantities) stems and leaves—alkaloids

Fruits

apples (seeds)	hydrogen cyanide
apricots (pits)	hydrogen cyanide amygdalin (produces cyanide on digestion)
cherries (pits)	hydrogen cyanide
lemons (seeds)	hydrogen cyanide
limes (seeds)	hydrogen cyanide
peaches	pits—amygdalin flesh—contains a goitrogen
pears	pits—hydrogen cyanide flesh—contains a goitrogen
rhubarb (leaves)	oxalate (causes corrosive gastroenteritis and affects serum calcium levels)
strawberries	contain a goitrogen

Nuts, Seeds, and Miscellaneous

almonds	hydrogen cyanide
bamboo shoots	hydrogen cyanide
beet sugar	saponin (enzyme inhibitor)
brown mustard	glycoside (a potent irritant)
cashews	contain a goitrogen
cocoa	caffeine (stimulant and teratogen) theobromine (stimulant)
coffee	caffeine
cottonseed (oil)	gossypol (pigment)
mace	myristacin (hallucinogen; toxic to liver)
nutmet	myristacin
peanuts	saponin contain a goitrogen
sugar cane	hydrogen cyanide
tea	green—saponin; caffeine black—caffeine; tannin (possibly carcinogenic); theophyline (stimulant)

Livestock Drugs

So you thought city streets were the only place with a drug problem? Visit a cattle feedlot or chicken hatchery. Some of the same antibiotics used in human therapy (penicillin and tetracycline, for example) are regular fare for beef and dairy cattle, poultry, and swine. They are used as much to boost growth as to treat disease. Although withdrawal times are required between last drug use and date of shipment for slaughter, these time periods may be insufficient protection for consumers of meat and dairy products. (See Chapter 7.)

Mould Toxins

Grains and nuts may harbour contamination by toxic moulds (fungi) due to poor harvesting and storage conditions. Unusual combinations of moisture and temperature are responsible for sporadic outbreaks of fungal toxins. Aflatoxin, a potent carcinogen produced by one of these moulds, is the chief cause for concern. (See Chapter 8.)

Pollutants

If a toxic chemical is released into the air or water, there's a good chance that some amount of it will end up in food. Some pollutants are widespread in occurrence. For example, it is estimated that fully one-quarter of the agricultural land in the United States is affected by lead from the gasoline exhaust of vehicles on the nation's highways, biways and sideroads. Other pollutants exert their effects only in a certain locale. Food contamination due to pollution from local industries occurs in communities across Canada, from lead in garden vegetables grown near Toronto lead smelters to mercury in Wabigoon River fish tainted by the emissions of the Dryden, Ontario pulp and paper mill. (See Chapters 9 and 10.)

Miscellaneous Food Contaminants

A survey of unintentional food additives can never be complete. New chemicals are constantly coming on the market; some of these may end up in food. Four types of miscellaneous food contaminants are discussed in Chapter 11.

Key Questions

As we proceed to look at environmental food contaminants in more detail, we'll be trying to answer the following questions:

- How toxic is the contaminant?

- What kind of toxic effects may be found?

- What levels of the contaminant are found in our food?

- What degree of risk do these levels of the contaminant present?

- What can be done to reduce the incidence of contamination?

- What, if anything, can consumers do to control their intake of each contaminant?

4 Controlling Contaminants in Food

IF THE PROBABILITY didn't exist that our food may be subjected to visits from toxic chemicals in our environment, there would be no need for the elaborate system at present in place to curb such contamination.

Canadian Laws on Food Contaminants

In Canada, conditions for proper handling of toxic chemicals that could wind up in the food supply are sprinkled throughout numerous federal and provincial statutes. At the federal level, there are, for example, the *Pest Control Products Act,* the *Feeds Act*, the *Grain Act*, the *Fisheries Act*, the *Clean Air Act*, the *Canada Water Act* and the *Environmental Contaminants Act*. But the final word on what is actually permitted in food, regardless of how it got there, is the *Food and Drugs Act* and the hefty set of Regulations that accompany it.

Section 4 of the *Food and Drugs Act* states outright that "no person shall sell an article of food that has in or upon it any poisonous or harmful substance" or "that is adulterated." At first glance, these statements would appear to provide the ultimate in consumer protection. But there are a couple of catches. The act does not spell out what constitutes "poisonous" or "harmful." Interpreting these terms is left up to scientists, industry, government, the courts, the consuming public. There will be disagreement about both which substances are poisonous (toxic) at all and what dosages of specific substances are harmful. So we're back to the debate about acceptable risk. "How safe is safe enough?" can be easily rephrased—"How harmful is harmful enough?"

However, what constitutes "adulterated" food *is* defined in Regulations under the Act, and these amount to exemptions from Section 4's edict against poisonous or harmful contaminants. That is, a specific food is considered adulterated only if it contains more than a specified level of a contaminant. This level is called a Maximum Residue Limit (M.R.L.), or tolerance. Substances for which tolerances have been established are shown in *Table 2*.

The rationale for setting tolerances is twofold. First, it is felt that even when great care is taken in agriculture and industry to avoid chemical contamination of food, a certain degree of contamination is unavoidable. If a few parts per billion of the carcinogen aflatoxin routinely taint peanut butter, the decision to be made is between

---

Table 2/ *Contaminants for Which Tolerances (Maximum Residue Limits) Exist Under Canada's Food and Drugs Regulations*

4   minerals (arsenic, fluoride, lead, tin)
90 pesticides*
9   livestock medications
1   mould toxin (aflatoxin)
1   pesticide metabolite (ethylene thiourea or ETU)
1   group of contaminants in pesticide manufacture (chlorinated dibenzo-p-dioxins)

* For some of these pesticides, tolerances include the pesticide's metabolites (breakdown products) as well. For example, DDT tolerances refer to "total DDT," i.e., DDT plus its metabolites DDD and DDE.

Source: Food and Drugs Act and Regulations.

---

prohibiting the sale of peanut butter altogether (which would not be a very popular move) and establishing a certain low level, a tolerance, which is attainable by proper growing, harvesting, and processing practices. Similarly, as long as we package fruit and vegetable products in lead-seamed tin cans, a small concentration of lead and tin will migrate into the food, in spite of all practical precautions taken in the manufacture and sealing of the cans.

The second rationale for tolerating a low level of contamination is that these levels are deemed toxicologically insignificant to humans. A tolerance for a certain chemical in a certain food product is always set such that normal human consumption of that food and other foods which may contain the same contaminant will not result in an intake greater than the Acceptable Daily Intake (A.D.I.) for that chemical.

The A.D.I., the amount that is without "appreciable risk" for humans when consumed over the course of a lifetime, is, in turn, based on the level of the chemical that has produced no observable effects in tests on the most sensitive species or strain of laboratory animal, that is, the No-Effect Level or N.E.L. A safety factor, typically 100, is placed between the No-Effect Level and the Acceptable Daily Intake. The A.D.I. is, therefore, at least 100 times more stringent than the N.E.L. The reason for throwing in a safety factor is that humans may be more sensitive to chemical contaminants than test animals; some individuals may be more susceptible than the average within the human population; and synergy with other chemicals in the diet may take place.

Acceptable Daily Intakes have been established internationally through joint efforts of the Food and Agriculture Organization (FAO) and the World Health Organization (WHO) of the United Nations (more about them later in this chapter). Different countries may adopt tolerances which are even lower than the level mandated by the A.D.I. if it is felt that such levels can be achieved routinely, regardless of toxicity. In Canada, tolerances are under a process of ongoing revision, as more information on health hazards becomes available. In general, the federal government adopts the FAO/WHO Acceptable Daily Intakes in determining tolerances.

But the Acceptable Daily Intake is not a scientific certainty, etched in stone, since the No-Effect Level may vary depending on experimental conditions. Furthermore, the hundredfold safety factor is entirely arbitrary. Some investigators feel that the A.D.I. may be appropriate for protecting humans against the "reversible", short-term health effects of certain contaminants. For example, the nervous system enzyme cholinesterase changes in function upon exposure to some pesticides; these changes "reverse" (i.e., stop) when exposure to the chemical stops. But these researchers feel that the A.D.I. is an unsatisfactory indicator of the cumulative, irreversible effects of genetically active chemicals—carcinogens, mutagens, and teratogens. So, while the concept of a tolerance, based on the A.D.I., is indeed a major advance in the evaluation of the safety of a food contaminant, and makes life easier for those charged with enforcing food quality laws, it is nevertheless a "guesstimate."

U.S. Food Contaminant Laws (A Thumbnail Sketch)

The U.S. equivalent to our *Food and Drugs Act* is the *Federal Food, Drug and Cosmetic Act* which, like its Canadian cousin, undergoes continual amendment. Portions of this act relating to environmental food additives are administered jointly by the Food and Drug Administration (FDA), the Environmental Protection Agency (EPA), and the United States Department of Agriculture (USDA). For example, the EPA holds the authority over the registration and banning of pesticides, a power which sometimes brings it into conflict with the USDA. The FDA polices the tolerances established for contaminant residues in food.

The International Framework

Advances in agricultural technology and improvements in long-distance transportation of food commodities in the past few decades have accelerated the pace of international trade in food. Standards for food quality, which vary among different nations, can become barriers to trade. In the early 1960s, some 135 agencies were working on international food standards and related problems. A pressing need surfaced for coordination of these efforts by a body of international stature. So, in 1963, two agencies of the United Nations—the Food and Agriculture Organization and the World Health Organization—established the Codex Alimentarius Commission to implement what became known as the Joint FAO/WHO Food Standards Program. The purposes of the program are given in the box on page 27. As of 1980, 112 nations had become members of the Codex Alimentarius Commission.

The FAO and WHO have a number of joint committees, such as the FAO/WHO Joint Expert Committee on Pesticide Residues, which are composed of toxicologists from several countries. After reviewing available scientific research, for example on agricultural use of pesticides and on pesticide toxicology, a joint committee proposes Acceptable Daily Intakes and Maximum Residue Limits for food contaminants. The appropriate Codex Committee, such as the Codex Committee on Pesticide Residues, made up of representatives of governments of major food exporting and importing nations, considers the FAO/WHO proposals and recommends Codex standards for residues in food moving in international trade. A standard, when adopted by a sufficient number of member countries of the FAO/WHO Food Standards Program, is published in a handbook called *Codex Alimentarius*. Once a country has agreed to accept a Codex standard, that country changes its legislation in order to enforce this standard. Canada does this by modifying the tolerances given in the *Food and Drugs Regulations*.

The ultimate goal of the Codex Alimentarius Commission is to establish worldwide acceptance of A.D.I.'s and tolerances for food in all member countries. While most industrialized nations have incorporated Codex standards into their food laws, the legal framework in many developing countries is not sophisticated enough to incorporate and enforce tolerances for food contaminants. However, in order to ship food commodities in international trade, developing countries informally adhere to Codex standards for food destined for export.

Objectives of the Joint FAO/WHO Food Standards Program

- To protect the health of consumers.
- To ensure fair practices in the food trade.
- To promote coordination of all food standards work undertaken by international governmental and non-governmental organizations.
- To determine priorities and initiate and guide the preparation of draft standards through and with the aid of appropriate organizations.
- To finalize standards and after acceptance by governments publish them in the *Codex Alimentarius* either as a regional or worldwide standard.

Of all the food consumed in Canada each year, 51 percent is produced domestically, 47 percent is imported from the United States, and 2 percent is imported from developing nations.

Acting on the Law: Policing Food

Analysis for environmental contaminants of food consumed in Canada and enforcement of the *Food and Drugs Regulations* on these chemicals are responsibilities of the Health Protection Branch (HPB) of Health and Welfare Canada. The Branch routinely samples raw (unprocessed) and processed foods, both domestic and imported, and performs appropriate tests at its five regional laboratories, located in Halifax, Montreal, Toronto, Winnipeg and Vancouver, as well as at a central laboratory in Ottawa. Testing for drug residues in animal products is done by Agriculture Canada as part of its meat inspection program and is coordinated with the activities of the HPB.

Two types of food inspection are carried out. Monitoring consists of sampling food products on a statistically random basis in order to get a picture of average levels of contamination. The more common form of inspection is surveillance, which zeroes in on known or suspected problem areas, such as a specific crop from a certain geographical location where pesticide use has been heavy in the past. Upon uncovering an incidence of contamination, Health Protection Branch inspectors may issue warnings to producers or processors to clean up their act or order product seizures. The HPB actually recalls products already on the market, if the degree of health hazard is judged to warrant such action. Imported products may be refused entry at border points. Prosecution is used only as a last resort. Records of seizures, recalls and prosecutions are published on a regular basis by the Branch, to inform food handlers of problems they might expect to encounter and to serve as a deterrent to the sale of contaminated products.

There is as much debate about the adequacy of current government control of food contaminants as there is about the chemical risks themselves. While producers and processors often complain about being "overregulated," some consumer interests maintain that the food sampling program is spread too thin and the enforcement procedures are too cumbersome to achieve sufficiently stiff control on toxins in food. The prudent path in controlling these health hazards would seem to be at least to keep up the present level of control. As the health effects of environmental contaminants become more clearly understood, further controls on their entry into the food chain will be in order.

PART TWO

The Additives

5 Pesticides

CRANBERRIES WERE IN short supply for Thanksgiving and Christmas of 1959 in much of the United States. Traces of the weed-killer (herbicide) aminotriazole, found to be carcinogenic in laboratory studies, contaminated some of that year's cranberry crop and led to its confiscation. Aminotriazole is still permitted for agricultural uses in Canada, although the residue remaining in food must be less than 0.1 parts per million (ppm). (In the United States, the herbicide is now used only for non-agricultural purposes.)

Twenty years later, in the autumn of 1979, five geese in British Columbia's Fraser Valley died after eating leaves from cauliflower being sold at a roadside vegetable stand. The actual cause of death was unclear, but infection or disease were ruled out. The cauliflower leaves contained residues of Monitor, an insecticide in widespread farm use in Canada to the tune of 2.4 parts per million, which is almost five times the allowable Canadian limit of 0.5 ppm. Questions were raised as to whether consumers could feel safe using cauliflower leaves in cooking.

Fortunately, the cranberry and cauliflower episodes are the exception rather than the rule. But they serve to indicate the potential for chemical risk-taking lurking in modern methods of growing food.

This chapter will begin with a discussion of agricultural pest control, the kinds of pesticides used, the ways in which pesticide use is regulated, and the environmental and human health effects of these chemicals.

33

Following that will be a description of how a pesticide ends up in a food product, then a look at the toxicity of the levels of pesticides found in food. The regulation of pesticides residues in Canadian food will be outlined, including the setting of tolerances. Then to the heart of the matter—information on what pesticides are actually showing up in our food and in what amounts, with special reference to meat, milk, and human breast milk. Trends over the past decade in total pesticide residues in a typical diet will be noted. After an examination of the problems of pesticides in products from Third World countries, the chapter will conclude with a look at how factory and home processing of food can reduce residues.

Of Pests and Pesticides

Pests are simply living things that happen to live where we don't want them. In the diverse, natural environment undisturbed by humans, a balance is achieved between plants and their pests—other plants (weeds), insects, rodents, fungi, viruses, and bacteria. But humans found that this natural balance wasn't good enough for intensive food production. As people domesticated plants and began to plant large areas with the same crop (monoculture), they produced an ecologically unstable but highly productive environment. These circumstances provide ample opportunity for the rapid build-up of pests, treated as they are to a seemingly endless source of nourishment. The pests' larder stays full, particularly if the same crop is grown on the same land year after year.

Worldwide, the total loss caused by crop pests in monocultures can approach 35 percent of the potential production. In order to prevent such loss in both the quantity and quality of food crops, we have introduced chemical pesticides into the environment. Although pesticides of various kinds have been used in agriculture since about A.D. 70, it has been only in the last 40 years that they have really come into their own.

Global production of pesticides now exceeds 1.8 billion kilograms (four billion pounds) annually, half of which is for agricultural uses. Indeed, as the *Canadian Medical Association Journal* put it: "Agriculture as we know it in North America could not exist without the widespread use of these important substances. By use of pesticides and other chemicals [such as synthetic fertilizers, as well as the use of newly-bred, high-yielding seed], Western agriculture has been able to produce unprecedented yields of food and fibre and has pointed the way towards at least a partial solution of the world food problem."

It has been estimated that in Canada there are some 2,800 species of insects, 250 species of weeds, 500 plant diseases (viral and bacterial), and 50 species of nematodes (a type of soil-dwelling worm) that are potential pests. Of these, about 300 insect pests and 150 weeds are considered to be of economic importance.

There are roughly 350 active ingredients used in the more than 16,000 pest control products registered for use in Canada. For the most part, the Canadian pesticide industry formulates (mixes active pesticidal ingredients with other substances that facilitate application) and packages pesticides. Active ingredients are imported from the United States, Europe, and Japan. Pesticide sales to farmers in 1977 amounted to over $160 million (see *Table 3*), or about 1.5 percent of the total farm cash receipts (income) of $10.1 billion for that year. Weeds are by far the largest pest problem and so weed-killers (herbicides) are the largest pesticide expense.

Table 3 / *Pesticide Sales to Canadian Farmers (1977)**

Type of Pesticide	Sales
Insecticides (for crops)	$ 21,131,935
Herbicides	125,159,352
Fungicides	7,944,823
Seed treatments	3,418,135
Growth regulants	851,749
Livestock insecticides	3,045,372
Rodenticides	742,373
Total	$162,293,739

* Latest figures available; this publication has been discontinued.

Source: Statistics Canada. *Sales of Pest Control Products by Canadian Registrants, 1977.* Catalogue 46-212.

Each class of pesticide—insecticides, herbicides, fungicides, and so on—can be subdivided according to the chemical structure of the compounds, which relates to the nature and degree of their toxicity.

Table 4 shows this breakdown, including common members of each subgroup and their acute toxicities.

Many of these synthetic organic (carbon-containing) chemicals were first developed for use in World War II. These include the · chlorine-containing (organochlorine) insecticide DDT and the phenoxy herbicide 2,4-D. Their inorganic predecessors, such as lead arsenate, had been less effective and highly toxic. (The presence of lead and other metals in pesticides is covered in Chapter 10.)

Table 4 / *Pesticide Groups and Subgroups and Their Acute Toxicities*

Group	Subgroup	Common Examples in Agricultural Usage	Acute Oral Toxicity (LD_{50} in mg/kg*)
Insecticides	Organochlorine	DDT	113
		chlordane	335
		methoxychlor	6,000
		endosulfan	43
	Organophosphorus	parathion	10
		malathion	1,375
		diazinon	108
		dimethoate	215
	Carbamate	carbaryl	400–850
		methomyl	17–26
	Botanical	pyrethrins	over 1,800
		rotenone	132
Herbicides	Phenoxy	2,4-D	300–1,000
		2,4,5-T	300
		MCPA	700–1,000
	Carbamate	barban	600
		triallate	1,675–2,165
	Dinitrophenol	dinoseb	5–60
	Substituted urea	diuron	3,400
	Triazine	atrazine	3,080
		aminotriazole	14,700
	Pyridylium	paraquat	157–207
		diquat	400–440
	Trichloro acid	trichloroacetic acid (TCA)	5,000
	Cyclohexene, phthalimide, and proprionic	dalapon	3,860–9,000

Fungicides	Guanidine and napthoquinone	dichlone	1,500
	Cyclohexene, phthalimide, and proprionic	captan folpet	9,000 10,000
	Ethylenebisdithio-carbamate (EBDC)	zineb maneb	over 5,200 6,750

* LD_{50} is the minimum dose of the pesticide, taken by mouth, in milligrams per kilogram of body weight of laboratory rats, which kills 50 percent of the animals. The *lower* the LD_{50}, the *more toxic* the pesticide. Pesticides are generally about four times more toxic orally than dermally (skin contact).

Important Note: These figures are indications of acute (short-term) toxicity. They are *not* a reliable indication of chronic (long-term) effects, such as cancer, which are not immediately observable. Note, for example, that the herbicide aminotriazole, of carcinogenic cranberry fame, has a relatively high LD_{50} of 14,700 mg/kg.

Adapted from:
Ontario Ministry of the Environment. *Pesticides Safety Handbook.* October, 1979.
Ontario Ministry of Agriculture and Food. *1979 Fruit Production Recommendations, 1979 Vegetable Production Recommendations, 1979 Field Crop Recommendations,* and *1980 Guide to Chemical Weed   Control.*

While the wartime herbicides are still in common use, the organochlorine insecticides were largely replaced in the early 1970s by insecticides belonging to two groups called organophosphorus and carbamate chemicals. By 1972, Canada banned DDT use on most food crops; very limited use is still permitted in some provinces.

While the chlorinated insecticides were cheap to use and had low acute toxicity to workers applying them, they are "broad-spectrum" (killing not only the "target" insect pest, but other, beneficial insects as well) and persistent, the latter characteristic leading to accumulation of the chemicals in the fatty tissues of crops, animals, and humans. By contrast, the newer organophosphorus and carbamate insecticides are more selective (that is, they display "narrow-spectrum" toxicity) and degrade quickly, largely eliminating problems of toxic residue in food. However, great care must be taken in their application because of high acute toxicity. Furthermore, since they are not persistent, they must be applied more often.

Organochlorine usage continues to be widespread in Third World countries, partly because of the expense and sophisticated application procedures involved in the use of organophosphorus and carbamate insecticides. (One of the heaviest uses of DDT globally is a non-agricultural one—the control of insect-borne diseases, especially malaria, typhus, yellow fever, and encephalitis.)

It is questionable whether the move toward the organophosphates and carbamates and away from the organochlorines is really an improvement. Also, since the organochlorines are so persistent and so mobile in the global environment (remember the DDT found in Antarctic penguins in the 1960s?), it is questionable whether organochlorine bans in industrialized countries have had much impact overall.

Agriculture Canada asserts that "we have probably a more sophisticated system of pesticide regulation than any other country in the world." All pesticides used in Canada must be registered for use under provisions of the federal *Pest Control Products Act and Regulations*. (The equivalent U.S. statute is the *Federal Environmental Pesticide Control Act*.) Section 3 of the Act is the ultimate authority regarding pesticide registration: "No person shall manufacture, store, display, distribute or use any control product under unsafe conditions."

All permitted compounds and usages are listed in the *Compendium of Pest Control Products Registered for Use in Canada*. A pesticide manufacturer applying for registration of a new product must produce detailed data that back up claims for effectiveness, toxicity, and environmental impact. With the onus placed on the company rather than some independent testing body, the potential for abuse is always present.

The data requirements are outlined in the *Pest Control Products Regulations*, as well as in trade memoranda issued periodically by Agriculture Canada. Research and red tape leading to registration may take from five to seven years and cost from $12 to $15 million. The odds against a compound originally synthesized and presented for screening reaching the final stages of commercial development as a registered pesticide are as high as 10,000:1.

Consumer confidence in this procedure (whereby pesticide manufacturers are responsible for providing toxicity data in order to register a pesticide) was shaken by a fiasco which first came to light in 1977 and again reared its ugly head in 1980. In 1977, Industrial Biotest Laboratories, an independent firm in Northbrook, Illinois that performs safety tests on pesticides for manufacturers, was found to have fixed much of the toxicity data, allowing pesticides to meet government standards for registration in both the United States and Canada.

Three years of intensive review of the work done by IBT revealed, in 1980, that two-thirds of the 405 studies done by IBT were in some way invalid. A further 410 studies have still to be reassessed. Fourteen of the 106 pesticides so far reviewed have been declared safe.

None of the pesticides in question have yet to be banned in Canada. States Dr. Alex Morrison, head of the Health Protection Branch of Health and Welfare Canada: "To make that kind of recommendation without evidence that it's needed, is going to disrupt the production of food in North America in a gigantic kind of way. And that has real impacts in terms of food costs, in terms of food availability, in terms of health.... This situation certainly can shake your assurance in the company involved. It is not a common practice, I can assure you, for companies to submit falsified data. In fact the pesticide manufacturers involved we don't believe knew about the data being falsified or fiddled with, or distorted. They were as chagrined by this as any of us were. The situation appears to have been related to problems at IBT. It is not a common problem in the whole of the industry, nor is it a common problem in the testing of chemicals in a generic sense. This was something which went wrong in a particular company."

Still, it is very disturbing to discover than as many as 36 pesticide manufacturers are believed to have relied on IBT for toxicity tests.

The Certificate of Registration for a pesticide product expires after five years. In most cases, the registration is renewed automatically. But the reregistration process does provide for cancellation (whereby existing stocks must be withdrawn and safely disposed of) or suspension (allowing existing stocks to be used up) of a registration if new data can successfully challenge the original data on which registration was based. So, as the Canadian Federation of Agriculture stated in its 1973 report on pest control in Canada, "a product is never finally approved."

Provincial governments are responsible for ensuring that pesticides are applied safely and within the restrictions imposed by the federal *Pest Control Products Act.* Provincial agriculture ministries issue annual pesticide use recommendations for farmers. Provinces may also choose to enact their own pesticide legislation, but such laws must be at least as stringent as the federal Act.

The benefits of pesticides cited earlier must not be viewed in isolation from the very real drawbacks they pose. Notes Clayton Switzer, dean of the Ontario Agricultural College at Guelph, "The use of pesticides has caused more unfavourable public reaction than any other agricultural practice." What is the basis of this backlash?

● **Diminishing Effectiveness**

Although harvest yields have almost doubled and the use of insecticides has grown tenfold since 1945, insects eat, each year, nearly twice as much of our crops now as then. The reason: Populations of insects, weeds, and fungi evolve genetically in favour of those organisms that are resistant to the killing effects of pesticides (see examples in *Table 5*). This phenomenon leads either to increased levels of pesticide application or to the search for and subsequent use of new, more effective (and more toxic) chemicals—the "pesticide treadmill."

Table 5 / *Examples of Pests' Resistance to Pesticides*

Type of Pesticide	*Specific Compound*	*Resistant Pest*
Insecticide	parathion	onion maggots
Herbicide	atrazine	weeds such as lamb's-quarters, common ragweed and redroot pigweed
Fungicide	benomyl	brown rot (a fungal disease of stone fruits such as peaches and cherries)

● **The Birds, the Bees, and the Drift**

Public uneasiness about pesticides in the 1960s focussed on the dramatic effects that the organochlorine pesticides, especially DDT, were found to have on the reproductive abilities of wild birds. Birds appear to be far more susceptible to the organochlorines than humans and other mammals. These effects on wildlife prompted bans on the organochlorines in most industrialized countries around 1970. The switch to organophosphates and carbamates saved the birds, but is taking its toll on the bees. Honeybees, which are necessary for pollinating many food crops, are extremely sensitive to the newer insecticides, such as carbaryl (Sevin) and carbofuran (Furadan). Pesticides destroy about 10 percent of U.S. honeybee hives and significantly reduce the populations of another 30 percent.

Both the birds and the bees are cases of "non-target" species suffering the toxic effects of insecticides. In general, herbicides are more specific in their action, since they are used to control plants (weeds) growing among other plants (crops). However, the drift of herbicides from fields of herbicide-insensitive crops (containing sensitive weeds) to sensitive crops is a serious problem. For example, 2,4-D may drift from areas planted with cereal grains to stands of vegetable crops, causing widespread damage to the latter. The toxicity to mammals of the common herbicides is relatively low.

● **People Problems**
While food processing additives seldom set out to kill (the exception being the antibacterial and antifungal preservatives), pesticides, by definition, are toxic to some living things and therefore can be suspected of being toxic to humans. The health effects of the organochlorine insecticides, if perceptible, would be due to residues of these persistent chemicals in food. By contrast, the main health threat posed by the organophosphorous and carbamate insecticides are health effects on people who apply these non-persistent but highly toxic newer chemicals. However, the organochlorines live on...and on, and although they lost their prominence in Western agriculture over a decade ago, their persistence makes them a continuing concern with respect to food residues.

Though only indirectly related to environmental contamination of food, occupational exposure to pesticides is of considerable importance in any discussion of their human health effects. *Appendix A* looks at farm worker exposure to these chemicals. The remainder of this chapter is devoted to pesticides in food.

How a Pesticide Becomes a Residue

A pesticide may be applied to a crop in one of two ways. It may be sprayed onto the aboveground, leafy part of the plant, in which case it can remain on the surface (to prevent mould growth in the case of a fungicide, for example), or be transported throughout the plant (as is the case with some insecticides). Alternately, a pesticide may be applied to the soil, absorbed through the root system, and carried to other parts of the plant; some herbicides work this way. A pesticide residue that is found only on the surface of the plant is called a topical residue, while one which has been distributed in internal tissues is a systemic residue.

The residue on or in the plant decreases over time, the rate of decrease depending on the chemical characteristics of each pesticide, particularly its stability or persistence. Some of the topical residue may be lost by evaporation and by the action of wind and rain. Systemic residues undergo chemical reactions in the plant that convert them to breakdown products that are usually non-toxic (exceptions are given in *Table 6*). Residues of pesticides applied early in the life of the plant will decrease in concentration as the plant grows and the residue is distributed throughout a larger volume of plant tissue. In a few cases, a pesticide may not be applied until after the crop has been harvested. Malathion, for example, is an insecticide sometimes used in stored grain.

Table 6 / *The Exception Rather Than the Rule: Conversion of Pesticide Residues to More Toxic Breakdown Products*

Pesticide	Breakdown Product (metabolite)	Type of Toxicity
parathion (insecticide)	paraoxon	increased inhibition of the nervous system enzyme cholinesterase
diazinon (insecticide)	diazoxon	as above
atrazine (herbicide)	several metabolites	carcinogenicity from extracts of atrazine-treated corn plants
ethylenebisdithiocarbamates or EBDCs (fungicides) such as maneb, mancozeb, and zineb	ethylenethiourea (ETU)	carcinogenicity in laboratory animals

Apart from residues of pesticides deliberately applied to a crop in a given year, additional residues may be found either as a result of pesticide drift from an adjoining field, or due to persistence in the soil of pesticides used in past years on the same field. Produce billed as "organically grown" at the retail store may accidentally contain pesticide residues through no fault of the farmer, but rather, because of drift from a treated field.

The focus of attention regarding persistent pesticides in soil has been the organochlorine insecticides. Both DDT and dieldrin have half-lives in soil of 8 to 10 years in Canada. (Organochlorine half-lives are shorter in tropical climates.) So even though organochlorine use has plummeted over the last decade, residues will probably continue to show up in food, although at progressively decreasing levels, for years to come.

It is commonly believed that the newer insecticides, especially the organophosphates and carbamates, as well as most herbicides, degrade so quickly that persistence in soil and water, and therefore possibly in food, is a non-issue. However, the matter is not that simple:

● While the second-generation insecticides are, indeed, less persistent than DDT and other organochlorines, they *can* leave soil residues, particularly when the soil is high in organic matter (humus). The organophosphate insecticide ethion is best known in this regard, and parathion has been known, on occasion, to persist in soil for up to 16 years.

● Residues of 2,4-D, the most commonly used herbicide in the Canadian Prairie provinces and previously thought to degrade completely within two weeks of application, have recently been found to occur year-round in Saskatchewan's Qu'Appelle River system and Manitoba's Red River system, both of which run through crop land.

Toxicity of Residues in Food

By far the largest proportion of pesticide exposure in the general population (that is, in people not occupationally exposed) is from residues in food, as indicated in *Table 7*. The intake of small amounts of pesticides in food over the course of a lifetime constitutes a body burden of foreign chemicals which may best be described as "chronic micro-insults."

American physician Wayland Hayes, Jr., in his comprehensive *Toxicology of Pesticides,* maintains that there have never been any cases of overt poisoning traced to eating food treated properly with pesticides and harvested according to good agricultural practice. However, two other circumstances present themselves: observable food poisoning from occasional, accidental misuse of pesticides, in violation of recommended instructions for application; and effects on health from proper pesticide use, resulting in legally acceptable residues, which may not be easily attributable to any one environmental factor such as pesticides. Considering the difficulty that the birds

Table 7 / *Sources of Pesticide Exposure in the General Population*

Source	Average Annual Intake of Pesticides (milligrams)
Food	30.00 mg
Absorbants and inhalants (cosmetics, aerosols, clothing, house dust)	4.96 mg
Air	0.03 mg
Water	0.01 mg
Total	35.00 mg

Source: H.F. Kraybill, "Significance of Pesticide Residues in Foods in Relation to Total Environmental Stress." *Canadian Medical Association Journal,* January 25, 1969, pp. 204-215.

and the bees have had coping with these complex, synthetic chemicals, it is reasonable to suspect similar problems when human beings eat pesticides in their food.

As Douglas Costle, the administrator of the United States Environmental Protection Agency (EPA) in Washington, has put it, "In the past we willingly accepted claims that pesticides have no long-term effect on humans. Neither EPA nor industry is in a position to make such reassurances honestly." In fact, the EPA has estimated that as high a proportion as one-quarter of the active ingredients in pesticides used in the United States are cancer-causing.

Concern that pesticide residues in food may be harmful to health is justified by the following findings:

• While many pesticides, such as the organophosphate insecticides, appear to be quickly metabolized and excreted from the human body, some of them, especially the organochlorine insecticides, overstay their welcome, persisting in body tissues for long periods. Some of the organochlorines are carcinogenic in laboratory studies.

• Research by Frank Duffy of the Harvard Medical School shows that human exposure to even minute amounts of some pesticides can alter brain activity for a year or more and cause irritability, insomnia, loss of libido, and reduced ability to concentrate.

● American allergist Granville Knight claims that pesticides can be responsible for flu-like symptoms and other cases of vague ill health.

● A growing number of individuals have been discovered to be unusually sensitive to pesticides and other chemicals in their food, and must go to great lengths to avoid the violent reactions that such exposure causes in them. This discovery has spawned a relatively new branch of medicine called clinical ecology, pioneered by Theron Randolph in the United States. A small but increasing number of physicians are recognizing and trying to treat these "ecological illnesses." In the forefront of this work in Canada is John Maclennan of Hamilton, Ontario.

● Pesticides may unwittingly contribute to a form of birth control. Male fertility in industrialized countries has dropped dramatically in the past 50 years and toxic substances such as polychlorinated biphenyls (PCBs) and pesticides are implicated. Ralph Dougherty of Florida State University notes that sperm density has declined from 90 million per millilitre of semen in 1929 to 60 million in 1979. Twenty-three percent of his recent volunteer squad of 132 college students were functionally sterile. *All* the samples contained pesticide residues.

● Pesticides may cause unpredicted, heightened toxic reactions in the body when ingested together with certain drugs and food ingredients. Organophosphate insecticides are more toxic in animals being treated with the drug phenobarbital, and carbamate insecticides may react with nitrites in food to form carcinogenic nitrosamines.

Policing Pesticides in Canadian Food

When a pesticide manufacturer seeks Canadian registration of a new control chemical, the manufacturer requests that a certain residue be allowed to remain in the food, consistent with scientific data supporting the inevitability and safety of that level of pesticide. Health and Welfare Canada advises Agriculture Canada as to whether or not a requested tolerance should be granted.

Pesticide tolerances were first entered into the *Food and Drugs Regulations* in 1956. Improvements in the sensitivity and specificity of techniques of laboratory analysis, such as the development of gas-liquid chromatography and mass spectroscopy, have allowed refinements to be made to tolerance levels. (Reliability of analytical techniques is sometimes still in question, for example with regard to carbamate insecticides such as carbaryl and carbofuran, and herbicides such as 2,4-D and atrazine.) Canada now has tolerances for 90 of the roughly 350 active pesticide ingredients used in agriculture.

Tolerances for some common pesticides are shown in *Table 8*. Note that the tolerance for a given pesticide may vary from food to food.

Table 8 / *Maximum Residue Limits (Tolerances) in Food for Some Common Agricultural Pesticides Under Canada's Food and Drugs Act and Regulations*

Pesticide (common name)	Maximum Residue Limit (ppm)	Foods
aldrin and dieldrin*	0.2 (calculated on the fat content)[+]	meat, meat by-products, and fat of cattle, goats, hogs, poultry, and sheep
	0.1 (calculated on the fat content)[+]	butter, cheese, milk, and other dairy products
benomyl, carben-dazin, and thio-phanate-methyl*	10.0	citrus fruits, peaches
	6.0	blackberries, boysenberries, raspberries
	5.0	apples, apricots, carrots, cherries, grapes, mushrooms, pears, plums, strawberries
	2.5	tomatoes
	1.0	beans, pineapples (edible pulp)
	0.5	cucumbers, melons, pumpkins, squash
captan	40	apricots, celery, cherries, grapes, leeks, lettuce, mangoes, onions, peaches, plums, shallots, spinach
	25	apples, avocados, beans, black-berries, blueberries, citrus fruits, crab apples, cranberries, cucumbers, dewberries, egg-plants, garlic, loganberries, melons, pears, peppers, pimentos, pineapples, pota-toes, pumpkins, quinces, rasp-berries, rhubarb, strawberries, squash, tomatoes

Pesticide (common name)	Maximum Residue Limit (ppm)	Foods
	2	almonds, beets, broccoli, Brussels sprouts, cabbage, carrots, cauliflower, collards, kale, mustard greens, peas, rutabagas, soybeans, sweet corn, turnips
carbaryl	10.0	apricots, asparagus, beet tops, blackberries, boysenberries, cherries, Chinese cabbage, citrus fruits, collards, cranberries, dandelions, dewberries, endive, kale, lettuce, loganberries, mustard greens, nuts (whole in shells), okra, olives (raw), parsley, peaches, plums, raspberries, salsify tops, spinach, Swiss chard, turnip tops, watercress
	7.0	blueberries, strawberries
	5.0	apples, bananas, beans, beet roots, broccoli, Brussels sprouts, cabbages, carrots, cauliflower, celery, eggplants, grapes, horseradish, kohlrabi, parsnips, pears, peas, peppers, poultry meat, radishes, salsify roots, tomatoes, turnip roots
	3.0	cucumbers, melons, pumpkins, squash
	2.0	barley, oats, rye, wheat
	1.0	corn, nuts (shelled)
	0.2	potatoes
carbofuran	0.5	carrots, peppers, potatoes, turnips (rutabagas)
	0.4	strawberries
	0.3	onions
carbofuran phenolic metabolites	1.0	carrots, potatoes, turnips
	0.5	onions, peppers, strawberries
chlordane	0.1 (calculated on the fat content)[+]	butter, cheese, milk, and other dairy products; meat and meat by-products of cattle, goats, hogs, poultry and sheep
2,4-D	5	asparagus
	2	citrus fruits

Pesticide (common name)	Maximum Residue Limit (ppm)	Foods
DDT, DDD, DDE*	7.0 (calculated on the fat content)[+]	meat, meat by-products, and fat of cattle, hogs, and sheep
	5.0	fish
	3.5	apples, celery, pears
	1.0	apricots, artichokes, asparagus, avocados, beans, beets, beet greens, blueberries, broccoli, Brussels sprouts, cabbages, caneberries, carrots, cauliflower, cherries, citrus fruits, collards, corn, cranberries, cucumbers, currants, eggplants, endive, grapes, gooseberries, guava, kale, kohlrabi, lettuce, mangoes, melons, mushrooms, mustard greens, okra, onions, papayas, parsnips, peaches, peanuts, peas, peppers, pineapple, plums, pumpkins, quinces, radishes, rhubarb, spinach, squash, strawberries, sweet potatoes, Swiss chard, tomatoes, turnips (rutabagas)
	1.0 (calculated on the fat content)[+]	butter, cheese, milk, and other dairy products; meat and meat by-products of poultry
	0.5	eggs
diazinon	0.75	apples, apricots, beets, broccoli, cabbage, carrots, cauliflower, celery, cherries, endive, grapes, kale, kohlrabi, lettuce, onions, pears, peppers, plums, salsify, spinach, strawberries, tomatoes, turnips (tops)
	0.70	citrus fruits, peaches
	0.50	beans, Brussels sprouts, cucumbers, turnips
	0.25	cantaloupes, collards, cranberries, figs, hops, lima beans, muskmelons, parsley, parsnips, radishes, summer squash, Swiss chard, watermelons, winter squash
endrin	0.02 (calculated on the fat content)[+]	butter, cheese, milk, and other dairy products

Pesticide (common name)	Maximum Residue Limit (ppm)	Foods
ethion	2.5 (calculated on the fat content)[+]	meat, meat by-products, and fat of cattle
	2.0	apples, citrus fruits, grapes, pears
	1.0	beans, peaches, plums, strawberries
	0.5	tomatoes
heptachlor, hepta-chlor expoxide	0.2 (calculated on the fat content)[+]	meat, meat by-products, and fat of cattle, goats, hogs, poultry, and sheep
	0.1 (calculated on the fat content)[+]	butter, cheese, milk, and other dairy products
malathion	8.0	apricots, avocados, blackberries, blueberries, boysenberries, cranberries, currants, dew-berries, gooseberries, grapes, loganberries, melons, mush-rooms, papayas, pecans, peppermints, pineapples, plums, raisins, raspberries, raw cereals, spearmint, strawberries
	6.0	asparagus, Brussels sprouts, cabbage, cherries, dandelions, endive, kale, lettuce, parsley, peaches, spinach, watercress
	3.0	cucumbers, leeks, lentils, okra, onions (green), pumpkins, salsify, shallots, squash, tomatoes
	2.0	apples, beans, pears, whole meal and flour from wheat and rye
	1.0	celery
	0.5	broccoli, cauliflower, collards, eggplants, kohlrabi, peas, peppers, Swiss chard, beets, carrots, garlic, horseradish, onions (dry), parsnips, potatoes, radishes, turnips (rutabagas)
parathion	0.7	beans, red beets, broccoli, Brussels sprouts, cabbage, carrots, cauliflower, celery, corn, cucumbers, eggplants, endive, kale, kohlrabi, lettuce, onions, parsnips, peas,

Pesticide (common name)	Maximum Residue Limit (ppm)	Foods
		peppers, pumpkins, radishes, spinach, squash, Swiss chard, tomatoes, turnips (rutabagas)
	1.0	apples, apricots, blackberries, blueberries, cherries, cranberries, currants, gooseberries, grapes, hops, loganberries, melons, peaches, pears, plums, quinces, raspberries, strawberries
toxaphene	7.0	beans, black-eyed peas, broccoli, Brussels sprouts, cabbage, cauliflower, celery, citrus fruits, eggplant, kohlrabi, lettuce, okra, onions, pears, peas, strawberries, tomatoes
	7.0 (calculated on the fat content)[+]	meat, meat by-products, and fat of cattle, goats, hogs
	5.0	barley, grain, sorghum, rice
	3.0	oats, pineapple, rye, wheat
	0.1 (calculated on the fat content)[+]	butter, cheese, milk, and other dairy products; meat and meat by-products of poultry

* Pesticides that are very closely related chemically.

[+] Since most pesticides concentrate in the fat portion of animal products, residue tolerances are based on pesticide concentrations in the fat rather than in the entire food. For example, DDT in milk fat is permitted at a level of 1.0 ppm; this may correspond to as little as, say, 0.01 ppm in the whole milk, since milk is only 1 to 3 percent fat.

Source: Extracted from Table II (Agricultural Chemicals), Division 15, *Food and Drugs Regulations* (Part B, Foods), p. 65A-65T. Current to June 1980.

Note: A number of the pesticides listed in this table are no longer used to any significant extent in Canadian or American agriculture. These include: aldrin/dieldrin, thiophanate-methyl, chlordane, DDT, endrin, ethion, heptachlor, toxaphene. However, they may still be present in foods imported from other countries.

These differences arise out of the varying tendencies of foods to retain residues, the lowest level of application that can be effective as a pesticide for each food, and the amount of the food that is found in the average person's diet (more about this last point later in this chapter). Tolerances are often granted for pesticides that have been virtually banned in Canadian agriculture—DDT, for example—since not only do residues of these pesticides often persist in soil, thereby continuing to contaminate crops grown in that soil and animals fed tainted crops, but also many of these pesticides are still permitted in countries from which Canada imports food.

No tolerances are given for over 250 of the 350 active pesticide ingredients. However, the *Food and Drugs Regulations* have recently been amended to include a clause which states that if no tolerance has been established for a pesticide, the allowable residue is automatically 0.1 part per million. This amendment makes life considerably easier for Health and Welfare enforcement officials. Without it, they had to prove, in any particular circumstance, that a pesticide without a tolerance was present in food at a "poisonous or harmful" level, as dictated in the *Food and Drugs Act*. The absence of a tolerance is an indication either that no residue above 0.1 ppm should occur under good agricultural practice, or that there is insufficient toxicological information available to establish an Acceptable Daily Intake and, from that, a tolerance. The 0.1 ppm blanket may be viewed as a set of interim tolerances, until more information on the toxicity of each pesticide becomes available.

The tolerances given in the Regulations apply not only to food grown and sold in Canada, but also to foods we export and import. The tolerance is useful in the case of foods imported from countries, particularly those of the Third World, which exert poorer control than Canada does on pesticide application in agriculture.

Farmers who follow provincially set recommendations for pesticide use should not exceed the federally established tolerances for residues in foods. The existence of a tolerance serves as a warning to farmers— that they will be in trouble if they misuse pesticides. One form that the recommendations for pesticide application take is the harvest interval, which is the time between last application of the pesticide and harvest. Examples are given in *Table 9*. (If Canadian farmers following pesticide application recommendations find that their products nevertheless contain illegal pesticide levels and therefore could be confiscated, they can file for compensation under the federal *Pesticide Residue Compensation Act*.)

The residue monitoring and law enforcement activities of Health and Welfare Canada seek to isolate cases where tolerances have been exceeded for pesticides having an established tolerance, and where the 0.1 ppm level has been overstepped for chemicals without tolerances.

Table 9 / Recommended Harvest Intervals for Some Pesticides Used on Ontario Fruit Crops

Pesticide	Harvest Interval (days)	Crops
captan (carbamate insecticide)	7	apple, pear, grape
	1	cherry, peach, nectarine, plum, blackberry, raspberry, strawberry, apricot
Carbaryl (carbamate insecticide)	7	apple, pear
	3	grape, apricot
	1	cherry, peach, nectarine
malathion (organophosphate insecticide)	7	apricot
	3	apple, cherry, plum, strawberry, currant, gooseberry
	1	blackberry, raspberry
methoxychlor (organochlorine insecticide)	21	peach, nectarine, apricot
	14	strawberry, currant, gooseberry
	7	cherry, plum
parathion (organophosphate insecticide)	14	apple, pear, cherry, peach, nectarine, plum, strawberry, apricot
zineb (EBDC fungicide)	30	cherry, grape
	14	apple
	7	pear, currant, gooseberry

Adapted from:
Ontario Ministry of Agriculture and Food. *1979 Fruit Production Recommendations.*

Roughly half of the produce tested is domestically grown and half is imported, with more seizures and confiscations of imported than domestic foods.

Under normal conditions of pesticide use, residues are rarely found which approach the established tolerances. This situation introduces a further safety margin, as does the fact that tests are performed on unwashed, untrimmed produce which might contain higher residues than the product actually consumed at the dinner table (see 'Cutting Down on Residues' later in this chapter).

Recent data from the Health Protection Branch indicate that no detectable residue is found in about 85 percent of samples tested, and that illegal residues are found in less than 3 percent of samples. But since only a very small percentage of all the produce consumed in

Canada is actually tested for pesticide residues, there is a good chance that foods containing excessive residues occasionally enter the marketplace.

Pesticide tolerances are in a constant state of refinement. They may be revoked when no longer necessary because of discontinued or changed use of the pesticide, and when new information shows that residues at the tolerance level could be a health hazard. Tolerances that have been revoked in the last few years include those for linuron, monuron and oxythioquinox (Morestan). Tolerances that have been reduced include those for chlordane, coumaphos, dalapon, ethion, and malathion. Health and Welfare Canada scientists have indicated that the tolerances for DDT and other pesticides long banned in Canada will likely soon be further reduced or even revoked, both because persistent residues are falling sharply and because we do not want to give foreign pesticide users the impression that Canada condones the use of these pesticides.

Observed Residues in Canadian Food: A Bird's Eye View

The first indications that the use of synthetic organic pesticides was not without complications came as early as the 1940s, when DDT residues began to show up in both animal and human tissues. The use of DDT and other organochlorines has taken a steady downturn over the past couple of decades and so have residues in food. But these pesticides still account for half of our dietary intake of pesticides. They show up almost exclusively in animal products, not because of direct application to animals but, instead, due to the phenomenon of bio-concentration (see 'Bio-Concentration' section later in this chapter).

By contrast, residues of the less persistent pesticides—organophosphate and carbamate insecticides, as well as most herbicides and fungicides—are found primarily in plant products. *Table 10* shows pesticide levels in Canadian food in the early 1970s. There were occasional occurrences in animal products of other organochlorine residues such as heptachlor epoxide, chlordane, and toxaphene. These three were also detected in plant products, though at much lower levels. The only plant material that has an inordinate ability to take up and retain organochlorines is carrot. Fourteen different organophos-

Table 10 / *Average Pesticide Content of Agricultural Commodities on the Canadian Market, 1972–1975*

Commodity   Pesticide Content in Parts per Billion (ppb)

Commodity	DDT* (OCl)	Dieldrin (OCl)	BHC⁺ (OCl)	Endosulfan (OCl)	Dicofol	Total Organo-phosphates
Milk and butter (fat)	34	15	16			
Beef (fat)	13	6	17			
Pork (fat)	44	1	3			
Poultry (fat)	60	12	5			
Eggs	5	<1				
Flour	1	<1	7			5
Potatoes	4	<1				<1
Carrots	39	4	1			5
Cabbage	<1	<1		3		
Beans	1		<1	1		
Tomatoes	1		1	1		
Vegetable oils	1	<1	6			43
Apples	1			1	5	112
Oranges	1	<1		<1	52	58

OCl organochlorine
<   less than (i.e., below limit of detection)
*   including metabolites DDE and TDE (DDD)
⁺   benzene hexachloride

Source: D.E. COFFIN, and W.P. MCKINLEY (Health and Welfare Canada). "Chemical Contaminants of Foods." *In: Proceedings of the Fifth International Congress of Food Science and Technology* (Kyoto, Japan, September, 1978). Amsterdam: Elsevier Scientific Publishing Co., 1979.

---

phates contribute to the 'Total Organophosphates' figures in the table. They consist primarily of malathion in flour (a postharvest grain treatment) and vegetable oils, diazinon in carrots, ethion and parathion in oranges, and phosalone and phosmet in apples.

The levels of organochlorine residues in dairy products for different regions of Canada are shown in *Table 11*. The food supply in the Prairie provinces of Manitoba, Saskatchewan, and Alberta is apparently less contaminated than that of any other region. The reason is unclear.

Some plant crops retain pesticides more than others. Leafy vegeta-

Table 11 / *Regional Distribution of Organochlorine Pesticide Residues in the Fat of Milk and Butter*

Region	Pesticide Content (parts per billion)	
	DDT	Dieldrin
Atlantic Provinces	32	12
Quebec	38	13
Ontario	35	28
Prairie Provinces	14	7
British Columbia	29	6

Source: D.E. COFFIN, and W.P. MCKINLEY (Health and Welfare Canada). Unpublished data, presented at the Fifth International Congress of Food Science and Technology, Kyoto, Japan, September, 1978.

bles such as broccoli, lettuce, and spinach have a large surface area relative to their volume, so can be expected to have higher topical residues. Legumes such as peas and beans (especially those in which the pod is discarded) have lower residues than other vegetables. Root vegetables concentrate residues in their outer tissues, which are often peeled away during food preparation, so that these vegetables end up with relatively low residues. "Garden vegetables" such as peppers, tomatoes, cucumbers, eggplant, and squash contribute more residue than other vegetables, as do most fruits. Residues in refined vegetable oils are much lower than those in the crude oils.

Although residues in domestic produce in excess of the legal tolerances under the *Food and Drugs Act* are rare, they do happen. For example, poor weather made 1977 a bad year for celery and lettuce in the Holland Marsh market gardening area north of Toronto. Unacceptably high levels of parathion and diazinon led to the confiscation of the highly perishable lettuce and storage of the celery until the residues reached safe levels.

Some common pesticides rarely, if ever, show up in food that reaches the supermarket. Since many herbicides are applied early in the growing season, Health and Welfare Canada has been hard-pressed to turn up any of the ubiquitous 2,4-D or the common corn herbicide atrazine. However, atrazine does persist in soil, so should be watched carefully in food.

Bio-Concentration: Pesticides in Meat and Milk

Although animal products make up only one-quarter of a normal
Canadian diet, half of all the pesticide residues are found in these
foods. This pattern is illustrated in *Figure 1*. The process by which
pesticide levels become magnified as they move up the ecological food
chain—bio-concentration—is responsible for this pattern of residues.
(Bio-concentration is described more fully in the Glossary; see Appendix B.) As well as this indirect process of animal food contamination,
there is the direct application of insecticides to livestock. For example,
insecticides are applied to the skin of dairy and beef cattle to ward off
warble flies.

Figure 1 / *Intake of Pesticide Residues in the Average Total
Daily Diet*

	Average Daily Intake (micrograms*)
Sugars and adjuncts	1.0%
Root vegetables	1.5%
Legume vegetables	2.0%
Potatoes	2.0%
Oils and fats	3.1%
Leafy vegetables	3.3%
Grains/Cereals	7.8%
Garden fruits	9.9%
Fruits	13.4%
Dairy products	14.4%
Meats, fish, and poultry	40.8%

5   10   15   20   25   30   35   40

Average Daily Intake (micrograms*)

* one microgram is 1/1,000,000 gram

Source: R.E. DUGGAN, and J.R. WEATHERWAX. "Dietary Intake of Pesticide
Chemicals." *Science,* vol. 157 (September, 1967), pp. 1006–1010.

Almost without exception, the pesticides found in meat and milk are the previously banned organochlorine insecticides. The reason is that they are the only pesticides persistent or stable enough to survive the bio-concentration process without breaking down into harmless by-products that pass out of the body. Since organochlorines have a passion for fats (lipids), they concentrate in the fat portion of animal food products. Therefore, the discovery of pesticides in a certain product, be it cheese, liver, or muscle meat, relates not to any intrinsic ability of that kind of product to hold pesticides, but rather, to the fat content of the food.

Dairy Products

Residues in dairy products are monitored intensively because of the importance of these foods, especially fluid milk, in the diet of infants, young children, and invalids. The Ontario government's pesticide residue testing laboratory in Guelph has been studying organochlorine residues in milk for the past decade. Laboratory director Richard Frank and his associates reported on milk collected in 1977 that while residues of DDT had decreased significantly from levels found in the early 1970s, dieldrin showed only a slight decrease, and heptachlor epoxide, chlordane, and endosulfan residues actually increased. Elevated levels of the latter two organochlorines can be attributed to the replacement of DDT by these pesticides when DDT was banned.

Most milk samples tested contained some detectable DDT, dieldrin, and heptachlor. But chlordane and endosulfan metabolites were found in only a few samples, and no residues of endrin, lindane, or methoxychlor were found in either the early or late 1970s. The Ontario data are summarized in *Tables 12* and *13*. Comparison between the levels in *Table 13* and permitted tolerances for dairy products shown in *Table 8* indicate that organochlorine residues in milk are only about one-hundredth of the maximum acceptable levels.

In the past, imported dairy products often contained excessive organochlorine residues. During the early 1970s, for example, some Italian and French cheeses coming into Canada were confiscated. These problems have abated since Italy and France and other European countries have banned or curtailed the use of many organochlorines.

Table 12 / *Incidence of Organochlorine Pesticides in Ontario Fluid Milk, 1977*

Pesticide or Metabolite	Incidence (% of samples)
DDE	97.1
Methoxychlor	0.0
Lindane	0.0
Dieldrin	99.3
Endrin	0.0
Heptachlor epoxide	99.3
Chlordane	4.2
Endosulfan or endosulfan sulphate	8.5

Adapted from: R. FRANK et al. "Organochlorine Insecticides and Industrial Pollutants in the Milk Supply of Southern Ontario, Canada—1977." *Journal of Food Protection,* vol. 42, no. 1 (January, 1979), pp. 31–37.

Table 13 / *Average Organochlorine Residues in Milk Fat in Ontario, 1977*

Pesticide or Metabolite	Mean Residue in Milk Fat (parts per million)
DDT	0.015
DDE	0.012
Dieldrin	0.011
Chlordane	<0.001
Heptachlor epoxide	0.004
Endosulfan	0.001

Adapted from: R. FRANK et al. "Organochlorine Insecticides and Industrial Pollutants in the Milk Supply of Southern Ontario, Canada—1977." *Journal of Food Protection,* fol. 42, no. 1 (January, 1979), pp. 31–37.

Meat

Organochlorine residues tend to be higher in meat than in milk simply because cattle hold onto their livers, tongues, and muscles longer than they hold onto their milk, giving the pesticides a longer time to bio-concentrate in the meat. Trends in DDT residues in Canadian meat from 1969 to 1974, monitored by Agriculture Canada's meat inspection division, are shown in *Table 14*. About 47 percent of the samples contained no detectable residue at all.

The reason that pork had DDT levels almost three times those found in beef is unclear. While DDT in beef, pork, and mutton gradually decreased over the five-year period, DDT in poultry apparently increased. The levels found are nevertheless only about one-hundredth of the tolerance of 7.0 ppm for beef, pork, and mutton and 1.0 ppm for poultry (*Table 8*). DDT levels in meat in the Prairie provinces were significantly lower than elsewhere in Canada, as was found to be the case for DDT levels in milk (*Table 11*).

Table 14 / *Trends in Residues of DDT in the Fat of Canadian Beef, Pork, Poultry, and Mutton, 1969–1974*

Time Period	DDT Residues (parts per million)			
	Beef	Pork	Poultry	Mutton
Oct. 1969–Apr. 1970	0.065	0.193	—	0.071
Apr. 1970–Oct. 1970	0.051	0.165	—	0.050
Oct. 1970–Apr. 1971	0.043	0.145	—	0.037
Apr. 1971–Apr. 1972	0.034	0.130	0.037	0.027
Oct. 1971–Apr. 1972	0.029	0.120	0.042	0.024
Apr. 1972–Oct. 1972	0.024	0.111	0.049	0.024
Oct. 1972–Apr. 1973	0.021	0.109	0.057	0.022
Apr. 1973–Oct. 1973	0.020	0.106	0.067	0.020
Oct. 1973–Apr. 1974	0.018	0.105	—	0.018
Apr. 1974–Oct. 1974	0.015	0.102	0.058	0.016

Adapted from: P.W. SASCHENBRECKER (Agriculture Canada). "Levels of Terminal Pesticide Residues in Canadian Meat." *Canadian Veterinary Journal*, vol. 17, no. 6 (June, 1976), pp. 158–163.

While sports fish usually come from freshwater lakes and rivers that are often polluted with DDT and dieldrin, commercial fish, usually of marine origin, are less contaminated. (The contamination of freshwater fish by DDT and other organic pollutants such as PCBs is discussed in Chapter 9.)

The Top of the Food Chain: People

Just as food animals store certain pesticides in their bodies, people eating animal products take bio-concentration one step further. The more animal fats a person eats, the higher will be his/her body burden of organochlorine pesticides, especially DDT, heptachlor epoxide, and dieldrin. Other organochlorines, such as methoxychlor, chlordane, endrin, and toxaphene are metabolized and excreted so efficiently that they are found only in the tissues of people who have had occupational or accidental overexposure.

Our bodies do not go on concentrating these chemicals ad infinitum. Rather, a balance is achieved between intake and excretion; that is, as long as the dosage remains the same, no more is stored in the body. The greatest residues are found in tissues of highest fat content. For example, DDT is stored to the greatest extent in adipose or fat tissue (the stuff that makes us overweight), to a lesser extent in the liver, and to the least extent in the kidneys, testes, and brain.

Average levels and incidence of organochlorine pesticides in the adipose tissues of Canadians in 1972 are shown in *Table 15*. *Everyone* has DDT, heptachlor epoxide, and dieldrin inside them! The levels found in 1972 were slightly lower than those found in a similar survey in 1969. But note how much higher DDT levels were in human tissue than in beef and other meats (*Table 14*)—bio-concentration strikes again.

Table 15 / *Organochlorine Pesticides in Adipose Tissues of Canadians in 1972*

Pesticide or Metabolite	Average Concentration (parts per million)	Percent of Samples Containing Residues
DDT and metabolites	2.571	100
Heptachlor epoxide	0.043	100
Dieldrin	0.069	100
Oxychlordane	0.055	97

Adapted from: JOS. MES et al (Health and Welfare Canada). "Polychlorinated Biphenyl and Organochlorine Pesticide Residues in Adipose Tissue of Canadians." *Bulletin of Environmental Contamination and Toxicology,* vol. 17, no. 2 (1977), pp. 196–203.

Because DDT and other organochlorines can, among other things, affect the nervous system, the question has been raised as to whether people experiencing dramatic weight loss might feel the effects of these pesticides as the individual's fatty tissues are "mobilized." Gordon J.

Stopps at the University of Toronto's Faculty of Medicine knows of no instances of overt human poisoning from pesticides as a result of weight reduction, but acknowledges that the possibility, though remote, is quite real. In *Toxicology of Pesticides,* Wayland Hayes, Jr. indicates that poisoning in rats previously fed large doses of DDT can occur when the rats are starved. However, human beings have a much lower metabolic rate than rats and so cannot lose weight, and therefore experience poisoning, at as fast a rate.

Because infants are often more susceptible to foreign chemicals than adults, we must consider what effects pesticides might have when they cross the placenta and contact the fetus, as well as when they are excreted in mothers' milk. DDT, a suspected teratogen, has been shown to be transferable from human mothers to their unborn. The transfer of DDT from mother to infant in the fat of breast milk is a further avenue for bio-concentration.

The levels of DDT and dieldrin in human breast milk in Canada are alarmingly high, as is shown in *Table 16.* The pesticides are much higher than in cow's milk, confirming that bio-concentration does take place. Note, too, that the breast milk levels exceed the Acceptable Daily Intake for DDT and dieldrin. The new mother, wishing to nurse her child, is faced with a thorny dilemma: Which does more harm,

---

Table 16 / *DDT and Dieldrin in Cow's Milk and Human Milk in Relation to Acceptable Levels, 1974–75*

Pesticide	*Daily Dietary Intake* (micrograms per kilogram of body weight)			*Acceptable Daily Intake* (micrograms per kilogram of body weight)
		Infant (1 month)		
	Adult	*Cow's Milk*	*Human Milk*	
DDT	0.028	0.20	6.60	5.00
Dieldrin	0.012	0.09	0.15	0.10

Source: D.E. COFFIN and W.P. MCKINLEY (Health and Welfare Canada). Unpublished data, presented at the Fifth International Congress of Food Science and Technology, Kyoto, Japan, September 1978.

depriving the infant of the nutritional benefits of breast milk or exposing her baby to high levels of potentially toxic pesticides in her milk? There is no simple answer.

(The presence in human milk of the industrial pollutants polychlorinated biphenyls (PCBs) will be covered in Chapter 9, as will the phenoxy herbicide contaminants called dioxins. Heavy metals in foods resulting from the application of metallic insecticides will be examined in Chapter 10.)

The Whole Ball of Wax: Total Diet Studies

In order to get a clear picture of the pesticide load borne by the average adult Canadian, Health and Welfare Canada conducts total diet studies. In such a study, foods comprising a typical Canadian diet are prepared for eating, but instead of sitting down to eat these typical meals, the laboratory analysts probe their pesticide contents.

Total diet studies were conducted annually from 1969 to 1973 and again in 1976–78 (one study). The food used in the studies was purchased at local supermarkets in the following locations: 1969—Ottawa; 1970—Vancouver; 1971—Halifax; 1972—Winnipeg; 1973—Toronto; and 1976–78—all five cities. In all cases, the levels of pesticides found were at consistently low levels—see *Table 17*. Total DDT intake decreased tenfold from 1969 to 1978, but DDT was still the heaviest residue by 1978. There has been no consistent downward trend in organophosphate residues.

Health and Welfare has discontinued the total diet studies. The reason? Residues found in past studies have been judged to be so low that continued analysis is not considered worth the effort. This reasoning can be explained by the relationship between residues found in the total diet and the Acceptable Daily Intakes as established by the Food and Agriculture Organization and the World Health Organization. Such a comparison can be made by examining *Table 17*. While in 1969, DDT was about 5 percent and dieldrin was over 50 percent of the A.D.I., in the 1976–78 study, DDT was only about 0.5 percent and dieldrin was about 2 percent of their respective A.D.I.'s. Dieldrin continues to be one of the few pesticides that is over 1 percent of its A.D.I. Others, including DDT, remain at less than 1 percent. These observations would seem to indicate that the tolerances given for unprocessed foods in the *Food and Drugs Regulations* are doing their

Table 17 / *Pesticide Residues in the Total Diet in Canada, 1969–1978*

Pesticide	A.D.I.* (micrograms per kilogram of body weight)	Daily Dietary Intake (micrograms per kilogram of body weight)					
		1969	1970	1971	1972	1973	1976–78
Organochlorine Insecticides							
DDT and metabolites	5.0	0.270	0.106	0.166	0.068	0.058	0.023
Dieldrin	0.1	0.057	0.016	0.032	0.021	0.025	0.002
Endosulfan	7.5	0.006	0.008	0.005	0.015	0.022	0.007
Endrin	0.2	—	—	0.005	—	0.002	0.001
Heptachlor epoxide	0.5	—	0.001	0.004	—	0.004	<0.001
Toxaphene	—	—	—	—	—	—	0.012
BHC	—	0.035	0.031	0.049	0.047	0.032	0.010
Organophosphorus Insecticides							
Diazinon	2.0	—	—	0.025	0.031	0.002	0.001
Ethion	5.0	—	0.019	—	—	0.004	<0.001
Malathion	20.0	—	0.030	0.042	0.010	0.009	0.012
Methidathion	—	—	—	—	—	—	0.012
Parathion	5.0	—	0.002	0.002	0.006	<0.001	0.003
Other							
Captan (fungicide)	100.0	—	0.024	—	—	—	0.004
HCB (fungicide)	0.6	—	—	—	<0.001	0.072	<0.001
Dicofol (miticide)	25.0	0.054	—	0.016	0.020	0.050	0.002
Chlorpropham (herbicide)	—	—	—	—	—	—	0.016
PCBs (industrial pollutants)	—	—	—	—	—	—	0.001

* Acceptable Daily Intake as established by the Food and Agriculture Organization and the World Health Organization

Source: H.A. MCLEOD, D.C. SMITH, and N. BLUMAN (Health and Welfare Canada). "Pesticide Residues in the Total Diet in Canada, V:1969 to 1978." *Journal of Food Safety,* 1980 (in press).

job of preventing pesticide intake in normally processed foods above internationally acceptable levels.

Similar studies, called Market Basket Analyses, have been carried out in the United States by the Food and Drug Administration, with similar results. The ten most prevalent residues uncovered in those studies are, in descending order of occurrence: DDT, DDE, dieldrin, TDE (or DDD), heptachlor epoxide, benzene hexachloride (BHC), endrin, aldrin, and toxaphene. All are organochlorines, with the first six accounting for 90 percent of the total intake.

However, the FDA has come under heavy fire recently for alleged inadequacies of the Market Basket surveys. Another U.S. government agency, the General Accounting Office (GAO) has been highly critical of the survey methods. The GAO found that residues of 195 of the 268 pesticides for which tolerances have been set in the U.S. are seldom checked. Examples include suspected carcinogens captan and dichlorvos, as well as the fungicides benomyl and maneb. The GAO also estimated that about 13 percent of all meat that goes to market may have pesticide residues in excess of established tolerances.

Other criticisms levelled against the Market Basket Analyses by the GAO include insufficient sample size and the practice of lumping similar foods together into composites, thereby obscuring the kind and amounts of residues that specific foods contribute. For example, Texas A & M University researcher Scott McKercher notes that while consumers would have to eat almost a kilogram (or two pounds) of raisins a day in order to exceed the A.D.I. of 0.125 milligrams of captan per kilogram of body weight, a medium-sized apple could easily provide a person with his/her A.D.I. for that fungicide. He went on to indicate that for most pesticides, there is at least one commodity that could cause excessive intake of the pesticide if the product contained residue at the tolerance level.

The U.S. Food and Drug Administration believes that the General Accounting Office has blown the problem out of proportion. Robert Chambers, who directed preparation of the GAO report, characterized the FDA and the Environmental Protection Agency (which sets pesticide tolerances) as "entrenched bureaucracies which don't like being told they're doing a poor job."

Presumably, the GAO might have similar criticisms of the Canadian government's earlier total diet studies, those covering the period from 1969 to 1973. Health and Welfare Canada's 1976–78 study is much more thorough than the previous studies in both sampling technique and number of pesticides monitored, but still retains some inadequacies shared with the U.S. Market Basket studies.

One shortcoming relates to what constitutes a typical diet. It is fine to determine pesticide tolerances for specific foods by using data based on the average or normal person's diet; but what about the millions of people whose diet is not normal according to the government's definition of the word? Such discrepancies were brought into focus early in 1980, when a group of health, environmental, and labour organizations in California sued the state's agriculture department for underestimating the consumer's pesticide exposure by using statistical averages to guess how much of any food is consumed by an individual.

The California suit was based on the Environmental Protection Agency's list of 81 foods that it considers to be seldom consumed, that is, not eaten in amounts greater than 210 grams (7.5 ounces) each *annually*. Included are such common items as avocados, blueberries, Brussels sprouts, eggplant, mushrooms, plums, rutabagas, and walnuts. The suit argues that the averaging technique does not realistically reflect the diets of many Americans. As an example, the EPA's method assumes that everyone eats fresh pears, when in fact perhaps 5 percent of the population actually does. If the latter is the case, then each consumer of pears consumes 2000 percent more pears than the EPA estimates he/she does.

Clearly, the methods by which tolerances are set and by which the total diet is determined merit serious reconsideration.

The validity of current tolerances may be brought further into question if the pesticide testing performed by Industrial Biotest Laboratories (mentioned earlier in this chapter) does, in fact, turn out to be fraudulent. Tolerances are arrived at from toxicities determined in such tests.

Pesticides in Third World Products

Over 90 percent of the produce imported into Canada comes from the United States, a country with a machinery for monitoring pesticide residues that is probably the most comprehensive in the world. (It does have its shortcomings, though, as was indicated above.) The remainder of the imported produce comes from Third World and other countries overseas, primarily Latin American nations.

One brand of conventional wisdom has it that since many pesticides long since banned in Canada are still used in developing nations and since these countries have very rudimentary control over pesticide use in local agriculture, produce imported from these places into Canada is loaded with toxic residues. Data obtained from the pesticide monitoring arm of Health and Welfare Canada for the three-year period of April 1977 to March 1980 do not bear out this wisdom, as is shown in *Table 18.*

Table 18 / *Pesticide Residues in Selected Imported Products, 1977–1980*

Country of Origin	Product	Tons Consumed in Canada, 1978 (from country of origin)	Number of Samples, Apr. 1977– Mar. 1980	Pesticide Residues
Chile	Onions	21,182	3	Nil
Honduras	Bananas[1]	12,181,900	—	—
Mexico	Tomatoes	9,371,000	48	Residues above tolerances in 3 samples.
Philippines	Pineapples[2]	431,163	2 (canned)	Nil
Trinidad	Sweet Potatoes[3]	751,361	—	—

[1]Bananas from other countries sampled in the same time period—Ecuador, Mexico, Panama—had no residues.

[2]Similarly, neither canned samples from Australia nor fresh samples from Hawaii examined in the same time period had any residues.

[3]Some sweet potatoes grown in the United States in this time period had trace residues, much below tolerances.

Sources: tonnages: Statistics Canada. *Agricultural Commodities,* 1978, vol. 1.
residue data: Field Operations Directorate, Health and Welfare Canada. Unpublished data, 1980.

The table does show, however, one major shortcoming of the Canadian monitoring program—a sheer lack of data. Over 12 billion kilograms (12 million tons) of bananas are imported annually from Honduras, yet Health and Welfare did not take a single sample from that source in three years. The health department openly admits that it does not have the resources to do much routine testing of imported produce. Rather, its efforts are biased towards products and countries that have been the source of problems in the past, such as tomatoes and peppers from Mexico. Also, its residue screening apparatus can detect only about 100 pesticides.

Health and Welfare is in the fortunate position of being able to fall back on the large volume of data accumulated by the U.S. Food and Drug Administration on both U.S. and Third World produce. The FDA also notifies Health and Welfare when produce with excessive residues might be headed for Canada.

In spite of the data given in *Table 18,* it is nevertheless reasonable to continue to suspect the existence of excessive residues in Third World produce, if only because of the relatively poor regulation of pesticides in those nations. Few developing countries have established legal tolerances for pesticides in food, applying, instead, the recommendations of the Codex Alimentarius Commission of the Food and Agriculture Organization/World Health Organization (see Chapter 4)— but often only for produce destined for export.

In other words, there is some effort made to meet FAO/WHO guidelines of acceptable residues, primarily because the importing country will simply refuse entry to produce contaminated with excessive residues. Needless to say, it is in their own best interests that producers in developing nations ship pesticide-free products. Lack of technological resources adequate to monitor crops for residues is likely the main roadblock to clean exports.

A serious problem remains, however, for the Third World countries themselves. Residues in crops destined for domestic consumption in countries with no legal limits on pesticides in food may be a significant factor in the diet. One survey of Central American farms found the organochlorine insecticide aldrin to be present at almost 2,000 times the level allowed in food in industrialized countries. The blood of people in Guatemala and Nicaragua contains DDT at an average of 30 times the U.S. level.

In late 1979, the U.S. magazine *Mother Jones* published a stunning exposé of the practice of dumping, whereby U.S. pesticide manufacturers ship to the Third World for use there, often under conditions that are hazardous to the health of farm workers, pesticides which have been banned or never even registered for use in the U.S. Like a boomerang, the illicit pesticides find their way back to the U.S. and Canada in produce we import from those countries.

Says *Mother Jones:* "A provision of U.S. law governing pesticides explicitly states that pesticides banned in the U.S. may be exported to other nations. The foreign buyers, in turn, either do not know or do not care that the chemicals have been found too dangerous for use in the U.S., the country where they are usually made....

"The 'vast majority' of the nearly one billion pounds of pesticides used each year in the Third World is applied to crops that are then exported *back* to the U.S. and other rich countries, according to WHO[World Health Organization]. This fact undercuts the industry's main argument defending pesticide dumping. 'We see nothing wrong with helping the hungry world eat,' is the way a Velsicol Chemical Company executive puts it. Yet, the entire dumping process bypasses the local population's need for food. The example in which DBCP [a pesticide which kills nematodes in soil and which is banned in the U.S. and Canada because it causes sterility] manufactured by Amvac [Chemical Corporation] is imported into Central America by Castle & Cooke Inc., [one of the largest foreign corporate landholders in Central America] to grow fruit destined for U.S. dinner tables is a case in point....

"The U.S. Food and Drug Administration reports that nearly half of the green coffee beans imported by this country are contaminated with pesticides that have been previously banned in the U.S."

If pleas for corporate social responsibility with respect to this kind of chemical dumping continue to fall on deaf ears, a tightening of pesticide manufacturing legislation in the U.S. and of tolerances for banned pesticides in both the U.S. and Canada will be in order.

Cutting Down on Residues

A careful distinction must be made between levels of pesticide residues found on raw, untrimmed, unwashed, unprocessed food and the food that we put in our mouths each day. In general, processing of raw agricultural commodities into food for the dinner table cuts residues substantially, adding a further safety factor to the one already used in arriving at pesticide tolerances under the *Food and Drugs Act*.

Most residue reduction is unintentional; it occurs as a (fortunate) by-product of normal processing procedures. It's not a question of the reduction being unanticipated; rather, it is just that it is incidental to the purpose of the processing method. For example, when crude vegetable oils are refined and deodorized by steam stripping at high temperatures, evaporation of organochlorine pesticides from the oil takes place. In this way, commercial processing of oils reduces organochlorine residue levels to below detection limits.

Trimming, Peeling, and Washing

Discarding the outer leaves of cabbage, cauliflower, and lettuce, the peels of bananas and oranges, the shells of nuts, and so on, reduces surfaces residues of pesticides. Similarly, peeling root vegetables (potatoes, carrots, turnips) and some fruits (apples, peaches, pears) reduces residues. There is, however, an unfortunate trade-off with peeling: While you are cutting down pesticides, you're also reducing the nutritional value of what you're eating, since vitamins and other "goodies" are often concentrated in the outer tissues of the fruit or vegetable.

But you can't go wrong with a thorough washing of all fresh produce. Some sources suggest either a dilute mixture of vinegar and water (to coax out alkaline-based pesticides), or soap or detergent (where feasible) plus thorough rinsing. But in some cases, detergent washing is actually less effective than plain water washing. Health and Welfare Canada officials feel that special washing procedures are probably not worth bothering about and instead recommend a simple washing in water.

No amount of trimming, peeling, or washing will remove systemic pesticide residues, those that have been taken into the interior of the food crop. However, the majority of residues in plant products are of a topical nature. (An example of the extent to which basic processing of tomatoes reduces residues is shown in *Table 19*.)

In animal products where the fat is localized, such as the fat at the edges of a cut of beef or pork, fat-loving pesticides can be largely avoided by trimming away the offending tissue. The solution is not as simple for dairy products, since the fat in whole milk, butter, and cheese is distributed throughout. A sensible precaution would be to cut down on the intake of these foods, not only to reduce pesticide intake, but to minimize exposure to saturated fats, which have been linked to vascular and cardiac disease. In some cases, an alternative exists—for example, skim milk and low-fat cheese.

Cooking and Other Processing

Cooking often reduces residues. Organophosphorous pesticides, which

Table 19 / *Pesticide Removal from Tomatoes During Commercial and Home Processing*

Pesticide	Original Residue (parts per million)	Percentage Loss of Original Residue			
		Washed	Washed + Detergent	Peeled	Canned Juice
DDT					
Commercial	7.7	89%	85%	99+%	99+%
Home	4.4	78%	—	99+%	99+%
Carbaryl					
Commercial	5.2	83%	96%	—	98%
Home	8.4	77%	—	92%	77%
Malathion					
Commercial	15.9	91%	83%	—	99+%

Source: JOSEPH C. STREET "Methods of Removal of Pesticide Residues." *Canadian Medical Association Journal,* January 25, 1969, pp. 154–160.

are not commonly found as residues in food anyway, are usually transformed into harmless compounds during cooking. Commercial canning of fruits and vegetables, which combines washing, peeling, and trimming with thermal processing, reduces organochlorines such as DDT substantially.

Pesticide reduction during the cooking of meat is variable. For example, cooked chicken has much lower DDT residues than raw chicken, but the same is apparently not true for beef. Regardless of the type of meat, longer cooking and discarding all fat drippings will contribute to pesticide removal.

The effect of milk processing on residues also varies. High-temperature production of milk powder removes much of the persistent organochlorine residue, through evaporation. By contrast, though, evaporated milk usually has a higher residue than the original milk. If milk has become contaminated with pesticides to the extent that it is unfit for consumption as is, it could be salvaged by processing it into whole-milk or skim-milk powder.

The milling of wheat redistributes residues; the refined flour has lower pesticide levels, but the remaining bran, the outer part of the kernel more accessible to sprayed pesticides, may have higher levels. However, whole-wheat bread aficionados will be pleased to learn that baking leads to some loss by evaporation of DDT, for example.

The only case in which processing unequivocally *increases* the residue problem is the conversion of ethylenebisdithiocarbamate (EBDC!) fungicides to a carcinogen and teratogen called ethylenethiourea or ETU during canning processes. While residues of these fungicides (e.g., maneb and zineb) have tolerances listed in the *Food and Drugs Regulations, no* detectable ETU whatsoever is permitted in food. Ian Munro, director of Health and Welfare Canada's Bureau of Chemical Safety, stated recently that "routine monitoring and surveillance of foods sold in Canada during the past few years has not detected any significant EBDC residues above existing residue limits. Although low residues of ETU were detected several years ago in some processed foods..., no significant residues of ETU have been found in foods during the past few years."

Nevertheless, Health and Welfare is keeping close tabs on EBDCs and ETU, permitting tolerances only for those crops for which no alternative fungicides are suitable under Canadian growing conditions and with a view to reducing the tolerances. But while the domestic scene may be under control, some imported products continue to harbour unacceptable residues.

A final few words about pesticides in food: With the recent increased interest in backyard vegetable and fruit production, attention must be given to the *proper* use of pesticides by persons who are untrained in their application and unfamiliar with the residue potential. While sound advice would be to read and follow *all* instructions given on the pesticide label, such advice should be immaterial. Why? Because no one needs to use toxic pesticides in a small-scale, diversified backyard garden. Consult one of the many handbooks of "organic" gardening and magazines such as *Harrowsmith* for clues to successfully grown, chemical-free crops.

What You Can Do

- Discard outer leaves, peels, shells of fresh fruits, vegetables and nuts, and consider peeling root vegetables (remembering the nutrient trade-off, though).

- Wash all fresh produce thoroughly in water.

• Remove fat from meat wherever possible and reduce consumption of high-fat dairy foods.

• New mothers—if you believe you have had undue exposure to pesticides by any route (food, air, water), you should have your breast milk tested. Insist that your health department perform this test (without charge).

• Grow your own! Many handbooks on no-pesticide (organic) gardening are available.

• Support all moves—by farmers, the farm industry, and government agricultural research agencies—towards ecological agriculture (see next chapter).

6 Alternatives to Pesticides: Ecological Agriculture

"SLOWLY BUT SURELY the stage is being set for a great national, indeed international, debate that will argue the wisdom of North American agriculture continuing indefinitely to move in the traditional direction of larger farms, fewer farmers, bigger tractors, more fertilizer, insecticides and specialization through monoculture, resulting in 'efficiency' and 'cheap food'."

Those are the words of Leonard Siemens, associate dean of the agriculture faculty at the University of Manitoba. Building on Siemens' opener is the view of McGill University agriculture professor Stuart Hill:

"During the last hundred years the North American food system has been transformed into a commercially efficient supplier of a diverse range of readily available, inexpensive and easy to prepare foods. This has largely been achieved by the breeding and selection of more productive cultivars [cultivated species] and varieties, and through the widespread application of increasingly powerful mechanical and chemical technologies to every phase of the food production and food handling process. The irony is that these very advances now pose a threat to the future stability of the system; firstly, because they have become increasingly dependent on resources and practices that are not sustainable; and secondly, because nourishment and fulfillment are not the primary goals within the system."

In this chapter, we'll examine why agriculture based heavily on chemical pesticides can't be sustained; what is meant by ecological or organic agriculture and whether this kind of farming will work; what is involved in Integrated Pest Management, a technique to reduce but not eliminate chemical pesticides; and, finally, what the political constraints are to wider implementation of ecological growing practices.

Chemical Agriculture

The current methods of growing our food—"chemical" or "conventional" agriculture—are a temporary arrangement. We can't grow on like this! In the jargon of the environmentalist, the technology is not appropriate.

Conventional agriculture has had a prolonged but, because of recent shortages, a slowly souring love affair with petroleum. Not only is this sort of agriculture dependent on oil and gas as an energy source and on gas-derived nitrogen as fertilizer, but also 80 percent of the pesticides used are petroleum derivatives. Carleton University geologist F.K. North has even gone so far as to say that we could define modern agriculture as the use of land to convert petroleum to food!

Conventional, petroleum-based agriculture is unsustainable, both because of fossil fuel shortages and because of the environmental hazards it creates; but it's tenacious. Notes Alvin Scheresky, a Saskatchewan farmer trying to implement ecologically appropriate practices: "Many farmers are beginning to question the long-term wisdom of intensive chemical use. [But] it's like a dope addiction. You don't know how to get out of it."

Many farmers, with the "help" of the pesticide manufacturers and the extension agents (farm advisers) of government agriculture departments, have come to see chemical pesticides as the only solution to pest problems. In many cases, there is overuse of these chemicals simply because information about the alternatives is not made readily available to farmers. They are encouraged to use pesticides as a form of crop insurance, spraying routinely in anticipation of a pest outbreak rather than in response to a specific problem.

The "tidy bug-hater" philosophy, as W.J. Turnock of the Agriculture Canada Research Station in Winnipeg puts it, pervades conventional agriculture. Good farmers, so the philosophy goes, do not allow *any* pests—insects, weeds, what have you—in their fields. Total eradication has been the name of the game, rather than simply controlling the level of pests below an economic threshold, the minimum level of pests than will cause significant crop damage.

The complex business of agricultural pest control may be somewhat simplified if we think in terms of chemistry and biology. Conventional pesticides are an attempt to solve what is essentially a biological problem—an imbalance, determined by economic considerations, between predator and prey (with respect to insects) or between competing plant species (with respect to weeds)—by using chemical (non-living, non-biological) techniques. These pesticides are a high-powered vehicle for working against natural processes, rather than augmenting already existing natural processes of pest control. A chemical insecticide, for example, often kills not only the target species but also that species' predators and parasites, thereby reducing the opportunities for natural and biological pest controls.

Carrying the argument over to the issue of crop fertilization, conventional agriculture has concentrated on chemically feeding the plants with specific ratios of key nutrients (nitrogen, phosphorus, and potassium), rather than maintaining a healthy and biologically active soil that in turn will more completely take care of the plants on its own, without the addition of chemical fertilizers. Conventional wisdom has it that soil is simply a physical support for the plant. By contrast, proponents of biologically based agriculture say that healthy soil makes for healthy plants and healthy plants resist pest attack.

In sum, then, conventional agriculture solves pest problems by treating symptoms chemically, one by one—what McGill's Stuart Hill calls "magic bullet" approaches—rather than looking further back to the causes of pest outbreaks and solving them biologically. Says Hill: "We do not suffer from pests because of a deficiency of pesticide in the environment just as we do not get a headache because of a deficiency of aspirin in the blood."

What is Ecological Agriculture?

To refer to biologically based agriculture as "organic" agriculture is somewhat misleading, so the term is best avoided. Originally, "organic" referred to any carbon-containing substance derived from living things. But the modern-day organic chemist studies all compounds of carbon, including those that never occur naturally in living organisms, such as the organochlorine insecticides like DDT. "Organic" has come full circle and is now often used to describe agriculture that employs only those fertilizers and pest controls that are derived from, or are themselves, living organisms.

There are actually several types of "alternative" agriculture, each with its own guidelines and often with its own philosophical basis. They include: organic or biological agriculture, a leading advocate of which is Rodale Press, publishers of *Organic Gardening* and *The New Farm* (formerly published jointly as *Organic Farming and Gardening*); bio-dynamic agriculture, a movement which is stronger in Europe and which is based on the theology and philosophy of Austrian philosopher Rudolph Steiner; the French intensive method, which concentrates on high productivity using small areas of land; and ecological agriculture, which is closest of the four to the mainstream of North American agriculture and which is espoused in the periodical *Acres U.S.A.*

These subgroups have several goals in common: stress on the quality of the food being produced and on practices which are in sympathy with natural processes, such as the substitution of biological for synthetic, chemical pesticides and the building of a healthy soil. For our purposes in the discussion to follow, the term "ecological agriculture" is most suitable and will refer to the general tenets of all the subgroups.

Daniel Zwerdling, an American journalist specializing in agricultural issues, has tossed out both "organic" and "ecological," preferring the term "sustainable agriculture." But the tenacity of the term "organic agriculture" means that it will crop up in spite of all efforts to bury it. For a useful, general definition of organic (or ecological or sustainable!) farming see the box on opposite page.

Can Ecological Agriculture Work?

Ecological agriculture has as many opponents as it has proponents. The common arguments proffered by its detractors, and the ecological responses, are as follows:

● *Ecological agriculture is a step backwards, back to inefficient, old-fashioned farming methods.*

Ecological agriculture does rely on some time-honoured practices such as crop rotation and tillage to reduce insect and weed problems, but it has also received a good dose of modern biology. It is not "pseudo-science" as those who misunderstand it might have us believe. As the U.S. Department of Agriculture's Study Team on Organic Farming has put it: "Most of today's organic farmers use modern farm machinery, recommended crop varieties, certified seed, sound livestock management, recommended soil and water conservation practices, and innovative methods of organic waste and residue management."

A Definition of Organic Farming

"Organic farming is a production system which avoids or largely excludes the use of synthetically compounded fertilizers, pesticides, growth regulators, and livestock feed additives. To the maximum extent feasible, organic farming systems rely upon crop rotations, crop residues, animal manures, legumes, green manures, off-farm organic wastes, mechanical cultivation, mineral-bearing rocks, and aspects of biological pest control to maintain soil productivity and tilth [soil structure], to supply plant nutrients, and to control insects, weeds and other pests."

Source: United States Department of Agriculture (USDA Study Team on Organic Farming). *Report and Recommendations on Organic Farming.* Washington, D.C.: U.S. Government Printing Office, July, 1980.

● *Ecological growing methods may be fine for the backyard hobby gardener, but they can never work on a large commercial scale.*

How, then, does one explain the success of large-scale ecological grain farms in the U.S. midwest Corn Belt? According to a landmark five-year study by the Center for the Biology of Natural Systems at Washington University in St. Louis, Missouri, farmers in Iowa, Nebraska, Minnesota and other states have been able to implement ecological practices and still flourish in the competitive farm economy of the Corn Belt (see *Table 21*).

● *Ecological pest control methods can't give enough protection against pests.*

One of the many examples of using insects to control other insects on a large scale comes from Gainsville, Florida. In the summer of 1975, entomologist R.I. Sailer released a parasitic wasp, *Pediobius foveolatus,* throughout Alachua County, Florida, to control Mexican bean beetles.

Sailer was so confident of the effect these parasites would have on the beetle population that he offered to *eat* all beetles found in the 1,000 square mile county during March and April of 1976. Despite efforts to provide Sailer with a steady diet, only 14 beetles turned up. The year before, three applications of insecticides were needed to control the beetles. The cost of Sailer's program was the salary of one graduate student for one year. In return, every bean farmer in the entire state of Florida was saved three years' worth of spraying costs.

Admittedly, ecological pest controls don't always work as well as Sailer's wasps. While the biological options for controlling insect pests are many and varied, there are fewer ecological avenues open for controlling weeds, fungi (plant diseases), and nematodes (a group of soil worms). Still, interest in "cultural" weed control (crop rotation, tilling, and plowing at appropriate times) has been rekindled, and there is talk of using chemicals given off by one plant species to halt the growth of other species.

Sometimes there are tricky trade-offs to be considered. For example, intensive tillage of land controls weeds biologically, but may lead to soil erosion, as well as using a lot of energy. But the alternative, energy-conserving application of chemical herbicides, means the possible build-up of resistance to the herbicide and the risk of toxic residues in food.

● *Ecological farming means drastically reduced yields per acre.*

A number of eminent sources use this approach to attack ecological agriculture. The Canadian Agricultural Chemicals Association, for example, estimates that our agricultural production could fall by one-third without the use of chemical pesticides. Similarly, former U.S. secretary of agriculture Earl Butz became famous (or infamous) for his claim that 50 million Americans would starve if we switched to organic growing methods.

These claims are unjustified for two reasons. First, a simple modification to one system of growing food, such as the withdrawal of pesticides from the conventional system without making any other changes (such as using pest-resistant crop strains), cannot give a valid picture of what can be achieved with another system. Second, no one is suggesting that we can switch over, whole hog, to ecological agriculture overnight. Since soil structure and composition, as well as naturally existing insect controls, may have been harmed by chemical pesticides, the change-over from conventional to ecological practices may have to be done gradually, over a period of several years.

A comparison of yields from organic and conventional farms is given in *Table 20.*

Table 20 / *Average Crop Yields on Organic and Conventional Farms in the Western U.S. Corn Belt, 1973 to 1976*

Crop	Bushels Per Acre	
	Conventional Farms	*Organic Farms*
Corn	78	75
Soybeans	28	32
Oats	47	64
Wheat	34	34

Source: United States Department of Agriculture (USDA Study Team on Organic Farming). *Report and Recommendations on Organic Farming.* Washington, D.C.: U.S. Government Printing Office, July, 1980.

● *Ecological agriculture is too costly; no one will pay the high prices necessitated by "going organic."*

A number of studies have indicated that ecological techniques cost the same as or only slightly more than chemical techniques. An example for the U.S. Corn Belt study is given in *Table 21.*

It is important to understand the reason why ecologically grown produce often does cost more at the retail level. As economist Robert Oelhaf points out in his 1978 book, *Organic Agriculture: Economic and Ecological Comparisons With Conventional Methods:* "[There is] a widespread belief that it is far more expensive to raise food organically than conventionally. This, however, is not a valid conclusion. Most of the differences have little to do with farm production costs. The main reasons for the differences are scale economies in processing, transportation, and retailing, and brand loyalty or quality differences, particularly at the processing stage....The point we wish to make is not that organic merchants are charging excessive amounts for their services. They may well be only covering their costs."

Table 21 / *Economic Performance of Crop Production on Ecological and Conventional Farms in the U.S. Corn Belt, 1974 to 1975 (Average of 14 Pairs of Farms)*

	Ecological	Conventional
Value of crops per acre of crop land	$164	$183
Operation costs per acre of crop land	$31	$50
Crop production returns per acre of crop land (value of crops minus operating costs)	$133	$133

Source: WILLIAM LOCKERETZ. "Can We Take the Chemicals Out of the Corn Belt?" *Horticulture,* September, 1977, pp. 14–16.

Presumably, as markets for ecologically grown produce expand through increased consumer demand, retail prices should fall. This prophecy has already come true in Switzerland, where the Migros supermarket chain works directly with Swiss farmers to provide ecologically grown food at competitive prices.

● *Ecological farmers are a bunch of off-beat, romantic dreamers.*

Quite the contrary. As Garth Youngberg of the U.S. Department of Agriculture's Study Team on Organic Farming put it, most organic farmers are "just hard-headed businessmen who feel they're doing better this way and will do better in the long run." Considering current farm economics, doing better may mean as little as simply breaking even.

● *Ecological agriculture is too complicated for the average farmer to handle.*

Farmers are not expected to be wellsprings of ecological knowledge. They can plug into extension services such as the "dial-a-bug" service for orchardists in Michigan and the "code-a-phone" service for vegetable growers in Ontario's Holland Marsh, which give telephone advice on weather patterns, insect populations and so on. These services are at present few and far between, but promise to multiply as interest in ecological methods grows. As British farmer Sam Mayall put it at a recent conference on ecological agriculture on Prince Edward Island, "I would say absolutely clearly that organic farming is viable for anyone who is prepared to meet its disciplines."

Organic agriculture was given an important boost in 1980 when the U.S. Department of Agriculture (USDA), not known previously for its support of alternative farming practices, issued a landmark report on the subject. A one-year study examined hundreds of organic farms across the United States, ranging in size from 4 to 615 hectares (10 to 1,500 acres).

The USDA's report concludes that organic farmers are farming more profitably and more productively than the agricultural (pro-chemical) establishment has ever believed possible. "Chemical farmers" in the U.S. use five times as much pesticide as they did only 15 years ago, but the USDA study team said it was "impressed" by the ability of organic farmers to control weeds and insects using natural methods. The 1980 report adds that the organic farms are "productive, efficient and well managed." Hundreds of thousands of small American farms could convert to organic methods in the next few years with little or no negative impact on the nation's economy, the study team found.

The fact that there is no universally recognized standard of ecological or organic farming makes for a ticklish marketing problem. Confusion about what is and is not an organically grown product runs rampant among both growers and processors eager to profit from jumping on the "clean food" bandwagon and consumers edgy about toxic residues in what they eat. One California food processor claimed that its applesauce was organic because the apples were mixed with distilled water instead of tap water. Another company insisted that its fruit was organically grown because chemical pesticides were used only before the fruit appeared on the trees.

Fraud is a problem in the organic food industry. To resolve this, local organizations in about half of the American states have set up certification programs for truly organic farmers. The governments of three states—California, Oregon, and Maine—and one U.S. federal regulatory agency, the Federal Trade Commission, have recently developed legal definitions of organically grown food. A key clause in the *California Organic Foods Act* of 1979 states that foods with pesticide residues in excess of 10 percent of the tolerances (maximum residue limits) set by the U.S. Food and Drug Administration may not be labelled as organically grown. (The 10 percent accounts for unintentional residues due to spray drift from "chemical" farms and general environmental contamination by pesticides, rather than direct application to crops.)

Similar moves to regulate and authenticate organically produced food are being made in Canada, though efforts here are not as well advanced as in the U.S. The Canadian Organic Certification Association and its affiliate, Greenleaf Whole Foods Ltd. of Kitchener, Ontario, are leading the way toward defining organic foods and implementing organic standards in this country.

At present, unless you know the farmer and his/her farming practices yourself, you really can't be sure your food has been grown without chemical pesticides, since taste and appearance may be identical to that of chemically raised products. As the USDA organic farming report notes: "Clear standards, it is argued, are essential to the growth of the organic foods industry. Many [organic farmers] believe that consumer confidence in organic foods depends upon the enforcement of strict certification requirements."

There is probably more ecologically grown food entering the market place than we know about. Some ecological farmers don't sell their crops to "health food" or "natural food" stores, where the small-scale distribution network means high prices. Instead, they channel their produce into the conventional marketing mechanism of the large agribusiness corporations, who mingle this produce with the chemically raised variety and sell it incognito at large supermarkets. For example, Nebraska farmer Del Akerlund sells his ecologically raised corn to CPC International, Inc., maker of Mazola corn oil. California grape farmer Stephen Pavich ships his organic table grapes to American supermarket chains like Safeway, Food Fair and Kroger.

Integrated Pest Management: The Least is Best Pesticide Strategy

The debate about which path for agriculture is "right," conventional or ecological, rages on. As economist Robert Oelhaf has said, "Both sides at times exaggerate or gloss over embarrassing facts." Robert Rodale of Rodale Press actually downplays the viability of ecological methods: "Many people have the impression that we're advocating a wholesale switch to organic farming, but that is definitely not true. We think that the arguments that organic farming can be done on a large scale are highly exaggerated and based on a very selective choice of facts. Our methods of composting and cultivation are not suitable for large corporate farms, and probably couldn't be done on a large scale."

What *can* be implemented on a large scale is Integrated Pest Management or IPM. IPM abandons chemical treatment as the first line of defence against pests, relying instead on ecological controls, and using chemical pesticides only if other control methods are ineffective. A detailed description, extracted from a periodical called *The IPM Practitioner,* is as follows:

WHAT IS IPM?

Integrated Pest Management is a process for determining WHEN you need treatment action (timing), WHERE you need treatment application (spot treatment), and WHAT strategy and mix of tactics to use. Strategies used may involve physical, cultural, biological or chemical controls.

In selecting tactics choose those that are:

(1) The least disruptive to naturally occurring controls upon the target pest population and other potential pest populations. (When you kill off the natural enemies of the pests, you inherit their work.)

(2) Most in harmony with both short and long term human and environmental health.

(3) Most likely to be relatively permanent.

(4) Easiest to carry out effectively.

(5) Most conserving of non-renewable energy fuels.

(6) Most cost effective in the short and long term (Use of a pesticide is no substitute for prevention).

An Integrated Pest Management program contains the following components:

(a) A MONITORING SYSTEM for regular sampling of the pest population and its natural enemies, potential pests in the same environment and their natural enemies, management decisions and practices that could affect the pest or potential pest populations, and weather.

(b) A determination of economic or aesthetic INJURY LEVEL—that size of pest population that can be correlated with an injury sufficient to warrant treatment of the problem.

(c) A determination of ACTION LEVEL: the pest population size, plus other variables such as weather, which allows prediction that injury levels will be reached within a certain time if no treatments are applied.

(d) An EVALUATION SYSTEM to determine outcome of treatment actions.

IPM has strong support from many pest control researchers. The United States Congress' Office of Technology Assessment has concluded that implementation of IPM for major U.S. agricultural crops "can cut pesticide use by as much as 75 percent in some cases, reduce preharvest pest-caused losses by 50 percent and save a significant amount of the one-third of the world's potential food harvest that is lost to all pests.... IPM is the most promising approach to U.S. crop protection."

Agriculture Canada likes IPM too. "We also regard IPM as a most promising method of pest control," says Elmer Hagley of the Vineland Research Station in Ontario. "In some fruit crops, pesticide use has been cut by 40 to 50 percent without adverse effects on crop quality. As far as acquiring the basic ecological data necessary for developing an IPM program is concerned, a period of 7 to 10 years is not unrealistic."

The pioneering work on Integrated Pest Management was done right here in Canada. In the 1940s, A.D. Pickett and his colleagues at the Kentville, N.S. Research Station developed an IPM program for Nova Scotia apple orchards. By careful manipulation of predator-prey relationships, the use of insect hormones (pheromones) and restricted use of narrow-spectrum insecticides, the east coast researchers have managed to control codling moths, oystershell scales, and European red mites on a crop which is notorious for pest damage. A remaining problem is scab; at present only chemical fungicides bring relief, but efforts are being made to breed scab-resistant apple varieties.

At Agriculture Canada's Research Station at Harrow, Ontario, an integrated control program for greenhouse pests has been worked out. Says the agriculture department's publication on the subject: "Programs have been developed that control whiteflies on cucumbers and tomatoes grown in greenhouses. An integrated schedule controls two-spotted spider mites and powdery mildew as well as whiteflies. The biological control agent, *Encarsia formosa* [a small wasp raised in Canada for pest control purposes as early as 1928], parasitizes the young stages of the whitefly. When the parasites are properly established, they keep whiteflies at low safe numbers for the rest of the growing season."

A program run by the Ontario Ministry of Agriculture and Food and the University of Guelph's Department of Environmental Biology uses Integrated Pest Management on 1,230 hectares (3,000 acres) planted with carrots and onions in the Holland Marsh and adjacent marshes. (This acreage represents about one-quarter of the total.) Guelph's Helen Liu, working out of the Muck Research Station in the marsh area, estimates that over the past three years, the use of insecticide sprays has dropped by 80 percent. Instead of spraying 10 to 12 times each growing season, onion and carrot growers in the study area are now spraying only up to 3 times and in some cases not at all.

IPM is not equally successful for all crops. It has been most successful with crops that are subject to almost annual attack by pests and had therefore been treated with chemicals routinely and frequently. Examples include potatoes, cotton, citrus fruits, and apples. Pest outbreaks that are sporadic due to their dependence on the proper weather conditions are less amenable to IPM. Grasshoppers in the Prairie provinces are a case in point.

As might be expected, proponents of ecological agriculture in its "pure" form (that is, *no* chemical pesticides) are somewhat critical of Integrated Pest Management. They maintain that IPM practitioners describing their work often tiptoe carefully around those situations in which pesticides *are* resorted to. We cannot assume that there will be no long-term health and environmental effects from even minimal usage of chemical pesticides.

McGill University's Stuart Hill has put it this way: "The move from purely chemical solutions to pest problems to IPM represents a move towards ecological agriculture. It is interesting to note, however, that most IPM programs that have been adopted are still relatively simple and usually include pesticides as a major component. The level of integration presently achieved is invariably much lower than that required to qualify as a sustainable system of pest control. The high powered, 'fix-it' strategies [i.e., 'magic bullets'] that presently dominate should be reserved for emergencies."

Implementing Alternative Agriculture: Political Constraints

If ecological agriculture and Integrated Pest Management hold so much promise as sustainable ways of dealing with pests, why are most farmers using *more* pesticides than ever?

"I'll give you some clues," says U.S. agriculture journalist Daniel Zwerdling.

"There are almost 3,000 special advisers in California who advise farmers on pest control. Eighty-five percent of these advisers are on the payrolls of pesticide companies. The vice-president of the massive University of California, home of the nation's most powerful agricultural school, sits on the board of Tenneco, the mighty agribusiness, oil, and pesticide conglomerate.

"Most of those farmers who have discovered that biological methods are a better way are afraid to say it out loud. Several farmers actually refused to let me use their names [in my reporting] because they say the huge, brand-name corporations they sell to will reject their crops if they don't use pesticides, so strong is the myth that only pesticides can control insects. It seems a crime of sorts, that a method of farming that protects the environment, conserves energy, and cuts farmer costs has to remain such a secret."

The influence of the pesticide manufacturers is not nearly as strong in Canada. Nevertheless, the U.S. situation affects us intimately if we remember that almost half of the produce we eat is imported from the United States.

What institutional changes will be required in order to implement alternative pest controls?

● First, ecological pest control information needs to be more readily accessible to farmers. The agricultural colleges which train people to act as extension agents to farmers and the governments which employ them still have a heavily pro-pesticide bias. Such influential people need to be convinced that ecological agriculture will work.

● Second, more research money needs to go to the development of ecological pest control strategies and the transition from conventional to ecologically based controls. Much of this expenditure will have to be public money. Saskatchewan's former agriculture minister Edgar Kaeding explains why: "Most research on controlling weeds and insects is done by corporations. They hope to develop a product that is effective, that can be patented and sold at a profit, thereby recouping their initial investment into research. We must remember, however, that they will not undertake research into *farming practices* that will control weeds and insects. This is because nothing can be patented and sold at a profit. As a result farmers, governments, or universities must be involved in undertaking such research."

Research needs identified by the U.S. Department of Agriculture's Study Team on Organic Farming are given below.

Research Recommended by
the USDA Study Team on Organic Farming

- Investigate organic farming systems using a holistic approach.
- Determine the factors responsible for decreased crop yields during the transition from conventional to organic farming systems.
- Determine the long-term effects on the productivity of selected soils from recommended applications of chemical fertilizers and pesticides in conventional farming systems.
- Develop efficient and safe methods for utilization of municipal wastes, especially sewage sludge*, as a source of plant nutrients and to improve soil productivity.
- Develop methods for more efficient recycling of nutrients in organic wastes for crop production.
- Determine the availability of phosphorus from rock phosphate and potassium from low solubility sources when applied to soils that are farmed organically.
- Develop refined soil test recommendations for nitrogen, phosphorus, and potassium based on crop, soil type, and associated climatic effects.
- Develop new and improved techniques for control of weeds, insects and plant diseases using biological nonchemical methods.
- Develop through breeding programs crop varieties that are adaptable to organic farming systems.
- Expand research on biological nitrogen fixation.
- Determine the effectiveness of organic wastes for improving the efficiency of chemical fertilizers.
- Conduct research on the potential impact of organic farming on the economic viability of small farms.
- Develop procedures to reduce use of antibiotics and chemicals in treatment of mastitis in cattle.**
- Conduct research to investigate the health safety of food products

exposed to, and possibly contaminated with, residues of pesticides and other synthetic chemicals in chemical-intensive farming.

● Conduct farm management studies to help individual organic farmers increase their incomes, and develop simulation or other type models to assess the aggregate socioeconomic impact of various combinations of conventional and organic agriculture.

Source: United States Department of Agriculture (USDA Study Team on Organic Farming). *Report and Recommendations on Organic Farming.* Washington, D.C.: U.S. Government Printing Office, July, 1980.

 * See Chapter 10.
** See Chapter 7.

Even ecological control practices that *can* be patented will be of little interest to large agribusiness corporations, since they are reluctant to invest large sums of money to develop narrow-spectrum, specific pesticides which, by definition, would have a limited market compared to conventional pesticides that are effective against a wide range of pests.

● Third, and also related to patenting, plant breeders' rights—the move to allow the patenting of new strains of crop plants—represent a serious threat to ecological agriculture and to the "gene pool" of plants well adapted to local environments. While breeding research can produce pest-resistant varieties, such as rust-resistant wheat, there is a widespread uneasiness that many large seed companies now controlled by oil and pesticide manufacturing conglomerates will go in the other direction, developing and patenting seeds which require the use of their own chemical pesticides.

● Fourth, we must reexamine the grading standards that have been set by farm products marketing boards and other agencies. Do we really need pretty, perfect produce? I spent the summer of 1972 in British Columbia's fruit-growing Okanagan Valley, where I enjoyed large volumes of peaches rejected at the packing house simply because of small bruises. Similarly, oranges in California are graded according to damage to the peel by citrus thrips, which have absolutely *no* effect on the taste or nutritional value of the part of the orange that we eat.

But lowering the grading standards will not be a simple matter. A Canadian Federation of Agriculture report on pest control in Canada has this to say:

"It has been suggested several times that some of the cosmetic aspects of the grading standards encourage and force producers to use pesticides, though these aspects do not in fact improve the quality of the product. For example, control of disease-causing organisms which cause the product to rot or of insect larvae which tunnel and destroy the fruit might justify fairly rigorous control procedures. Other aspects such as insect "stings" [or other] skin defects which do not really affect the quality of the product other than through its appearance are cosmetic uses of pesticides and are due to grading standards that perhaps should be reconsidered. Perhaps the highest quality grade could tolerate more insect stings, or skin defects.

"Another consideration might be that if the grading standards are adjusted so that [the product is] less attractive to the consumer, perhaps the market for that product would drop off and therefore an action which was taken to change grading standards based on environmental considerations could indirectly harm the production [and marketing] of that commodity."

● Fifth and last, we need to adopt a broader social outlook when it comes to defining what kind of agriculture is appropriate. As Daniel Zwerdling has said: "Obstacles to organic agriculture—and they are formidable—are less inherent in organic methods than they are products of narrow social vision. Like racehorses guided by blinders, the nation's policymakers cannot envision other worlds beyond the chemical and agribusiness track. There are solutions to these obstacles—but they would require bold national policies for social change.... What if some organic farms *do* require more labor? In a nation where millions remain unemployed and trapped in decaying urban ghettos, it is ironic to hear agriculture officials bragging about how few workers are employed on the farms."

So, while the debate over chemical versus alternative agriculture is far from over, let's give the ecologically based alternative a fighting chance!

What You Can Do

● Encourage Agriculture Canada and Consumer and Corporate Affairs Canada to introduce a certification program for organic farmers and provide a clear definition of what constitutes an organically grown product. Without a definition, you can't be confident that a health food store's "organic" produce is truly organic.

● Ask the proprietors of your health food store who their produce suppliers are, and find out more about these farmers.

● Your only avenue to ensuring that a product is grown without pesticides is to visit the farm and talk with the farmer.

7 The Barnyard
Medicine Chest

ONE EVENING IN 1979, a quick supper of hot dogs landed a Winnipeg teenager in hospital with a potentially fatal allergic reaction. Subsequent laboratory analysis showed that the hot dogs contained traces of penicillin, to which the boy was severely allergic. One man's meat had become the same man's poison.

This chapter is about the use of drugs in livestock production. It will discuss the antibiotics and hormones used to treat animal diseases and boost growth; the hazards inherent in using these drugs and in consuming meat and dairy products from treated animals; government regulation of animal drugs and residues of them in our food; the presence of penicillin in milk; and, finally, efforts to reduce the use of drugs in rearing livestock.

Fatter is Finer

How did one of modern medicine's most commonly prescribed antibiotics find its way into a frankfurter? It all began late in the 1940s, when the livestock industry was seeking ways to increase its capacity to produce meat and dairy products for a rapidly expanding postwar population. Just as crop farmers were looking at ways to increase yields, animal producers were searching out methods for increasing the productivity of swine, cattle, and poultry.

The idea of rearing large number of animals in confined spaces had achieved limited success until that time, largely because of the constant threat of infectious disease and associated stresses of crowded quarters. What permitted the transition to the "factory farming" techniques of cattle feedlots and poultry "batteries" was the discovery that antibiotics could be used not only to prevent and treat disease, but also, together with certain hormones, to stimulate faster growth and more efficient conversion of feed to flesh and fat. Thanks to a combination of better nutrition, sophisticated breeding, and growth-promoting drugs, a hen that produced 200 eggs annually in 1950 produced 250 eggs annually by 1970.

Intensive livestock production is probably here to stay. The perceived advantages are good returns to the producer, lower retail prices, uniform (not better) quality, and provision of animal protein to masses of people.

The disadvantages associated with overcrowded, automated livestock factories include inhumane treatment of the animals and lowered quality of food for the dinner table, especially increased fat and water content and the presence of drug residues. But these drawbacks have not yet figured high on the balance sheet.

So if you thought drugs were only for people, just open the barn door. Animal medications are a big business. Over one-third of all antibiotics sold in Canada each year are for livestock. Drugs are administered at some point in the life of most commercially raised food-producing animals in North America.

Drugs may be administered in drinking water or feed, they can be injected, or pellets can be implanted under the animal's skin. Addition to feed is the most common route for antibiotics; 80 percent of antibiotic application is through medicated feed. There are three uses for drugs in livestock production: to treat disease, to prevent disease, and to promote growth through increased weight gain and improved feed efficiency. Disease treatment with drugs, or therapeutic use, is often in the hands of a veterinarian, while the latter two drug uses, called subtherapeutic uses, are generally the responsibility of the livestock producer.

Antibiotic dosages used in feed depend on the intended effect. Disease treatment usually requires from 200 to 500 grams (7 to 18 ounces) of antibiotic per ton of feed, disease prevention from 50 to 200 grams (2 to 7 ounces), and growth promotion from 1 to 50 grams (0.04

to 2 ounces). The longer an animal is fed antibiotics, the larger the dose required to improve growth, sometimes as high as 200 grams per ton of feed.

While the way in which antibiotics work to treat disease is well understood—they kill off disease-causing (pathogenic) bacteria and other harmful micro-organisms—just how they work to promote animal growth is unclear. The most plausible theory is that the drugs suppress those organisms that cause low-level disease—disease which is without symptoms but which decreases weight gain nevertheless.

Antibiotics used commonly in food animal production include penicillin, tetracycline, and streptomycin. Not strictly antibiotics but having similar effects are the sulfa (sulfonamide) drugs. A third class of drugs are the nitrofurans. Finally, medications which can promote growth but not treat or prevent disease include copper and arsenic compounds and hormonal preparations of melengestrol acetate (MGA), Synovex (estradiol benzoate with progesterone or testosterone), and Ralgro (zearalenol). The hormone diethylstilbestrol (DES) was widely used in both feed and implants until its ban in 1974 in Canada and 1979 in the United States. *Table 22* indicates some specific drugs and common uses.

There is no question that these drugs are effective in boosting animal productivity, thereby providing economic benefits to the producer. Recent Canadian data indicate that with the efficiency of conversion of feed to animal tissue improved by, for example, 7 percent for beef cattle, 4 percent for swine, and 3 percent for broiler chickens when feed contains growth-promoting additives, the dollar value of the feed saved through this increased efficiency totals some $25 million annually. Furthermore, while decades of antibiotic use have, indeed, built up populations of drug-resistant bacteria, these drugs somehow continue to be effective in improving animal performance.

But—and this is a big and interesting but—most drugs administered to livestock have significant effect only when the animals live in a highly stressful environment. The factors which make antibiotics work best are manure, disease, unbalanced diet, extreme temperatures, and crowding—in other words, the common stresses of the factory farm.

Medication has been replacing sanitation and other practices contributing to the general well-being of farm animals. Notes William Buck, a toxicologist at Iowa State University: "We use too many drugs

Table 22 / *Drugs Commonly Used in Livestock Production*

Drug Type	Specific Compounds	Common Uses
penicillin	penicillin, ampicillin, cloxacillin	poultry primarily; also swine, dairy cattle
aminoglycosides	streptomycin	poultry, swine, cattle
tetracycline	oxytetracycline, chlortetracycline, tetracycline	all food animals
sulfa (sulfonamides)	sulfamethazine, sulfathiazole, sulfadimethoxine, sulfaethoxypyridazine, sulfaquinoxaline	young swine primarily, also cattle, poultry
nitrofurans	furazolidone, nitrofurazone	poultry primarily; also swine
copper*	copper sulphate	swine
arsenicals*	arsanilic acid, 3-nitro-4-hydroxyphenyl arsonic acid	swine, poultry
hormones	melengestrol acetate (MGA), Synovex (estradiol benzoate plus progesterone/testosterone), Ralgro (zearalenol)	cattle

* Livestock drugs containing metals are also discussed in Chapter 10.

in place of good management and sanitary practices. We could put our emphasis on a little better management and disease control instead of on drugs and do just about as well. Of course that would put some hardship on the farmer; it would require more physical labor to clean out the pens, and keep them washed better."

Hazards of Medicated Meat

Roy Atkinson, a former president of the National Farmers Union, has acknowledged that there are many Canadian livestock producers who refuse to eat the animal products they sell to the public. Such farmers keep separate animals for their own consumption, in order to avoid residues of drugs administered in the factory portion of their farm operations. Atkinson's admission is a telling indication that there may be serious risks to the consuming public in the way food animals are raised in North America.

The risks are of two kinds:

• Drug Residues

After administration of an antibiotic or hormone to an animal, most of the drug is ultimately either eliminated in the urine or feces, or broken down in the body into by-products (metabolites) that are biologically inactive. But in some cases, there may be storage of the drug or certain harmful metabolites in the animal's body tissues, which means that small amounts of the medication may remain in the edible portions that make it to the supermarket meat or dairy counter. As our Winnipeg teenager found out, these residues can pose a problem for individuals with allergies or other hypersensitive responses to certain antibiotics. Apart from allergic reactions, there is the concern that a few of the drugs appear to be carcinogenic. The hormone diethylstilbestrol (DES) was withdrawn from livestock use in Canada and the U.S. primarily because it was thought to contribute to adeno-carcinoma, a rare form of vaginal cancer. (The primary source of DES, however, was not residues in meat, but, rather, its administration during pregnancy as a means of preventing miscarriage. Daughters of women who had received DES during pregnancy have developed adenocarcinoma decades later.) Other animal drugs highly suspected or proven to cause cancer include the nitrofurans and Synovex-H (an estradiol benzoate/testosterone propionate combination used in heifers).

• Proliferation of Drug-Resistant Bacteria

The problem here is not at all one of toxic residue in food, but, rather, an insidious biological chain of events. When drug-sensitive bacteria in a farm animal's body are killed or inhibited by an antibacterial drug, drug-resistant bacteria, those few which are genetically adapted to the presence of the drug, are given the chance to flourish and become the majority. Furthermore, these resistant bacteria can, through mobile bits of genetic material called plasmids, transfer their resistance to other bacteria of the same or different species. This heightened resistance may have two results. First, it may compromise the effectiveness of the drug in livestock production, necessitating either higher doses or a switch to another antibacterial. Second, the resistance may be transferred to an until-then sensitive strain of a bacterial species which also causes disease in humans, such as those of the *Salmonella* group. This transfer of resistance may occur either in the animal's digestive tract or in that of a human who has contacted bacteria from animals or animal products. There then develops a reservoir of resistant bacteria in the human gut that can spread their resistance around and make antibacterial drug treatment ineffective. Some bacteria become resistant not to just one, but to several antibiotics at once.

The extent to which widespread antibiotic usage in rearing food animals actually spreads drug resistance in both animals and humans is not well understood. Many medical and veterinary researchers feel that the majority of antibiotic resistance in human bacterial populations is caused not by drug residues in animal products, but by liberal prescribing habits on the part of physicians. In Canada, antibiotics account for 15 to 20 percent of all prescriptions and 30 to 35 percent of hospital drug costs. A recent study of 188 patients at a Manitoba children's hospital, reported in the *Canadian Medical Association Journal,* found that antibiotics were prescribed unnecessarily in 13 percent of medical cases and 45 percent of surgical cases.

Regardless of the relative contributions of human and animal drug use to the resistance problem, the task remains to reduce such use wherever possible, since there have been a number of cases in which human bacterial infections, such as *Salmonella* food poisoning, have shown an unexpectedly poor response to antibiotic treatment.

The top priority at the food production end should be to eliminate low doses of antibiotics from animal feeds, since these constitute a continuous supply of drugs with which bacteria can work to develop resistance. In addition, continuous, low-level use for purposes of growth promotion reduces the effectiveness of the drugs when they are really needed for disease treatment.

There is considerable debate over which of the two hazards of medicated meat is the more critical and how serious either of them really is. The U.S. Congress Office of Technology Assessment, in its 1979 report *Drugs in Livestock Feed,* maintained that the health risks from the development of bacterial resistance to antibiotics in feed are of greater concern than the risks of cancer from, for example, the nitrofuran furazolidone. V.W. Hays and W.M. Muir of the University of Kentucky's Agricultural Experiment Station feel that the most pressing concern for tissue residues is with the sulfa drugs and for bacterial resistance, penicillin and tetracycline.

Policing the Medicine Chest

Charles Gracey, general manager of the Canadian Cattlemen's Association, has said of livestock farmers: "If we conduct ourselves properly and there are no drug residues, then the consuming public is placed at no risk." So the two elements of the regulatory scheme for preventing drug residues in animal products are the laws surrounding livestock drug use and the degree to which farmers adhere to them.

Regulations under the federal *Food and Drugs Act* prohibit the sale of a drug for use in animal production unless data on safety and usefulness of the drug have been reviewed and considered satisfactory by the Bureau of Veterinary Medicine of Health and Welfare Canada's Health Protection Branch. (In the United States, the Department of Agriculture and the Food and Drug Administration have joint regulatory authority over drugs in animal products. The key piece of legislation is the *Federal Food, Drug, and Cosmetic Act* and its Animal Drug Amendments of 1968.)

In Canada, it normally takes about five years from the filing of an Investigation of New Drug Application in Ottawa to the marketing of the product. Feeding trials and toxicity tests are conducted at farms, feedlots, and universities. The cost involved in having a new medication cleared may approach $1 million.

Inclusion of drugs in Canadian animal feeds is controlled by the *Feeds Act and Regulations.* All feeds must be registered under that Act, which is administered by the Plant Products Division of Agriculture Canada. The agriculture department publishes a *Compendium of Medicating Ingredients,* which contains detailed drug clearances, approved levels, drug compatibilities, length of treatment, and withholding periods (discussed in greater detail later in this chapter) for all medications approved by Health and Welfare Canada for use in Canadian animal feeds.

With very few exceptions, Canadian regulatory policy on drug residues in food of animal origin is that there be absolutely no detectable residues at all. This policy embodies the general prohibition of

Section 4 of the *Food and Drugs Act* against the sale of food "that has in or upon it any poisonous or harmful substance." The exceptions (see *Table 23*) constitute a retreat from Section 4, just as do pesticide tolerances. Furthermore, drug metabolites which may be present in animal products are not covered by the no-residue blanket.

Some animal products tend to favour drug retention more than others. High-fat products such as butter and cheese, and beef and chicken livers are prime target for attention in a no-residue policy. In addition, drug residues which do occur are often not destroyed by cooking or other processing. For example, penicillin residues in milk do not break down during pasteurization.

Enforcement of the no-residue regulation is performed by Health and Welfare Canada and Agriculture Canada at packing plants and egg grading stations. Any animal product found to contain drugs is condemned and confiscated. Drug residues in milk are monitored by provincial authorities.

Long-time federal Agriculture Minister Eugene Whelan has said on numerous occasions that "Canada has the best meat inspection system in the world." But the enforcement system certainly is not airtight. Notes David Campbell, director of HPB's Bureau of Veterinary Medicine: "Although monitoring and applying penalties is a fairly effective deterrent, it does not provide complete protection to consumers. It is impractical to monitor all carcasses. We must depend on the deterrent effect of random testing, especially at the packing plant and producer levels. Although we are committed to the no detectable residue policy, the high safety factors afford protection to consumers even if small drug residues accidentally occur."

Residues do, indeed, turn up, though their extent is unclear. In the spring of 1980, Member of Parliament and veterinarian Gus Mitges alleged that 14 percent of Canadian meat and poultry is contaminated with drug residues. But an in-depth study by Agriculture Canada carried out in 1977–78 on 4,200 cattle and hog carcasses put the figure at only 0.12 percent and further analyses now in progress are indicating a level below 1 percent. The agriculture department does acknowledge, though, that there still remains a problem with sulfa drug residues in pork.

The chief tool the livestock farmer has available to avoid drug residues in meat or dairy products is proper observance of the withdrawal period given for each drug. This is the time span required for

Table 23 /
Exceptions to Canada's No-Drug-Residue Policy

Drug	Tolerance (parts per million)	Foods
arsanilic acid	2.0 (as arsenic)	pork and poultry liver
	0.5 (as arsenic)	pork meat, poultry, meat, eggs
amprolium	0.5	uncooked muscle meat of poultry
	1.0	uncooked liver and kidney of poultry
	7.0	eggs
buquinolate (Bonaid)	0.1	poultry muscle
	0.4	poultry meat and by-products, poultry kidney and liver, poultry skin and underlying fat
clopidol (Coyden 25)	10.0	uncooked poultry tissue
	25.0	uncooked liver and kidney of poultry
crufomate (Ruelene)	1.0	meat and meat by-products
decoquinate (Decox)	1.0	poultry muscle
	3.0	poultry meat and by-products, poultry kidney and liver, poultry skin and underlying fat
3-nitro-4-hydroxy-phenylarsonic acid	2.0 (as arsenic)	pork and poultry liver
	0.5 (as arsenic)	pork meat, poultry meat, eggs
ronnel (fenchlorphos)	7.5	fat of cattle, sheep, goats
	3.0	fat of swine
zoalene	2.0	chicken fat
	3.0	chicken and turkey meat, turkey fat
	6.0	chicken kidney and liver, turkey kidney and liver

Source: Food and Drugs Regulations (Part B, Foods), Table III, Division 15, pp. 65U–65V (Current to June, 1980).

drug residues to disappear between last drug administration and shipment to market. The withdrawal period varies from hours to weeks depending on animal species, route of administration (feed, injection, implantation), dosage, persistence of the drug and its metabolites in the animal, and toxicity of residues. Withdrawal periods for most drugs are specified by the manufacturer and are given on drug or feed labels and in the *Compendium of Medicating Ingredients*. In a few cases, such as injection of penicillin for disease treatment in cattle, withdrawal periods are subject to direct federal control through regulations under the *Food and Drugs Act*. Some representative withdrawal periods are shown in *Table 24*.

Table 24 / *Some Withdrawal Periods for Livestock Drugs in Canada*

Species	Drug	Route of Administration	Withdrawal Period (days)
Chicken and turkey	furazolidone	feed	5
Chicken and turkey	amprolium	feed	0
Swine	oxytetracycline	injection	18
Swine	nitrofurans	feed	5
Swine	sulfonamides	feed	80
Cattle	dihydrostrepto-mycin	injection	30 (milk discard— 96 hrs.)
Cattle	estradiol benzoate progesterone	implant	70
Cattle	sulphamethazine	feed	10 (milk discard— 96 hrs.)

Source: Health and Welfare Canada. Unpublished data, October, 1979.

Market-place pressures are a primary reason that drug-contaminated animal products sometimes reach the dinner plate. As Health and Welfare Canada officials Ian Munro and Alex Morrison have put it: "Whether withdrawal periods are strictly adhered to is an area of concern for regulatory officials. It seems reasonable to assume that the time of slaughter of most production animals will coincide with peak market prices and as a result it is suspected, though by no means proven, that withdrawal periods may at times be abused to some extent by some producers."

Tied to the problem of withdrawal period observance is the fact that a higher and higher proportion of total livestock drug sales are being made over-the-counter directly to the farmer, sidestepping the veterinarian. Continuous use of medicated feeds is undoubtedly a cause for concern, but because feed mills mix the drugs into the feed in precise amounts (the feed's "guaranteed analysis") approved by Agriculture Canada, there is probably less room for flagrant abuse in the growth-promoting uses of drugs through feeds than there is in disease-treating uses. Faced with rising costs of production, a livestock farmer may try to diagnose and treat animal disease without incurring a vet bill. "I try a needle of penstrep [penicillin/streptomycin] and if that doesn't work, then I call the vet."

In a report on the health of food-producing animals commissioned for the government of Ontario in 1976, physician E.H. Botterell created some waves in the agricultural community, including antagonism from the agriculture minister himself, when he reported the following:

"Information has been sought regarding the use by farmers or their employees of 'designated' antibiotics, that is without veterinary advice or prescription and for livestock.... Designated antibiotics are used for treatment purposes indiscriminately, often inappropriately and excessively. The farmer may suffer financial loss from lack of veterinary advice regarding diagnosis and proper therapeutic use of antibiotics, and unnecessary expense. Increasing resistance of microorganisms has been seen in laboratories and in veterinary and human medical practice. The judgment is inescapable that indiscriminate use of antibiotics has contributed, and continues to contribute to the continuing development of resistant strains of microorganisms in animals and man. It is unfortunate that the increasingly thoughtful and selectively critical use of antibiotics by veterinarians and physicians is now accompanied by uncritical, non-selective use of designated antibiotics by farmers and their employees."

There may also be financial pressures on the veterinarian. Since vets sell drugs to farmers directly, rather than the farmer taking a prescription to a pharmacy to be filled as would be the medical equivalent, Botterell suggested that "the possibility exists of veterinarians being influenced consciously or sub-consciously in the direction of overtreating and overprescribing because of fiscal gain from the sale of drugs. The profits to the veterinarian from the sale of all the livestock medicines needed by a farmer for intensive production of food animals will exceed, in some instances by a substantial margin, the professional income earned by the veterinarian, on an hourly rate."

Unintentional contamination of feed with improper levels of drugs can happen, though it is uncommon. In 1975, hog feed mixed by a small Canadian company contained about ten times as much growth-promoting antibiotic as the company specified in its guaranteed analysis. Another route for contamination may take place on the farm. If a feed bin containing feed treated with a drug is later used for untreated (clean) feed, the latter can become contaminated. So the farmer can be feeding what he/she thinks is clean feed to older animals that are almost ready for market without realizing that the feed may harbour drug residues.

A final market-place pressure which compromises the federal no-residue policy relates to the perishability of most animal products. Just as tests on food crops for pesticide residues may not be completed before the fruits and vegetables have gone to market, enforcement officials often cannot analyze and recall meat or dairy products containing illegal drug residues before they are sold and consumed. Sometimes all that can be done is to isolate the source of the contamination and prevent future shipments into the market place from that source until the problem is resolved.

Milk

Let's zero in on one key animal food product to get a closer picture of drug residue contamination. Milk may or may not be "the perfect food," but it does figure prominently in the diets of Canadian youngsters, and dairy products are certainly prominent in most of our diets.

Dairy cattle are plagued by mastitis, a bacterial infection which causes inflammation of the udder. About 50 percent of the Canadian dairy herd is infected in an average of two quarters of the udder. Milking machines are the major reason for the incidence and persistence of mastitis. Defective machines and improper operation irritate the cow's mammary tissues, promoting infection and allowing the spread of existing inflammation. Other factors contributing to mastitis incidence include unsanitary conditions and stress.

Mastitis costs Canadian dairy producers over $500 million annually in direct production losses (that is, reduced milk output) and an additional $200 million from animal death and culling, discarded milk, drugs, and veterinary expenses. The gain in milk production through controlling mastitis clearly outweighs the control costs. Control involves a long-range management program, consisting of proper attention to herd health, careful maintenance and operation of milking machines, good personal hygiene during milking, treating of teats with disinfecting agents and skin conditioners, and treatment with antibiotics, usually penicillin.

The withdrawal period following antibiotic treatment of dairy cattle is much shorter than for beef cattle, since residues are more persistent in muscle tissue. Depending on the drug and formulation, the period is in the order of 72 to 96 hours for milk.

Milk is tested routinely at the dairy for the presence of antibiotic residue. Ninety percent of residues found are penicillin; occasionally tetracycline and chloramphenicol show up. Incidence of detectable residues is low, which is good news for the more than 150,000 Canadians allergic to one or more antibiotics. For example, in 1978 in Ontario, 148,736 milk samples, involving 430 producers, were tested, of which 458 samples, or about 0.3 percent, were positive for antibiotics. But whenever tainted milk is found, a fine is slapped on the producer. In Ontario, it amounts to 60¢ per hectolitre (100 litres), increasing to $1.20 per hectolitre if the milk is still positive on subsequent retesting. The milk is dumped until negative results are obtained.

In the summer of 1979, the government of Quebec, Canada's largest dairy-producing province, instituted an on-farm milk testing program that eliminates the problem of test results becoming available only after the milk is in the market place. The Quebec program, which uses a mobile testing laboratory that can give results within a couple of hours, allows inspectors to dump tainted milk before it is shipped to the dairy and processed.

In the United States as in Canada, the federal government enforces a zero tolerance for penicillin residue in milk. The dumping of penicillin-contaminated milk in the U.S. is supervised at the level of the dairy cooperatives. If the tainting of a truckload of milk can be traced to a particular farmer, he or she is required to pay compensation for the entire truckload. This may consist of 500,000 to 1,500,000 gallons of milk.

Dairy producers pay close attention to antibiotic residues in milk.

Fines for contaminated milk are a reasonably effective deterrent to antibiotic abuse. Besides, it is in the farmers' best interest to produce antibiotic-free milk, since dairy processors who buy their milk cannot produce cultured foods such as cheese, buttermilk, yogurt, and sour cream from drug-tainted milk. The bacteria required for the manufacture of these products cannot grow in such milk.

Apart from and perhaps more serious than residues in milk, resistance to antibiotics is known to occur among mastitis-causing bacteria, for example in the species *Staphylococcus aureus*. More attention must be paid, then, to mastitis prevention and methods of mastitis control other than drug treatment.

Weaning from the Walking Drugstore

The potential human health effects of the use of drugs in animal husbandry were appreciated as early as the 1950s. One of the first official recognitions of the problem was the report of a committee chaired by M.M. Swann to the government of Great Britain in 1969 (the so-called Swann report). Swann emphasized the threat from bacterial resistance to antibiotics, and received considerable criticism in the press from various quarters, including scientists who felt the evidence for the development of resistant strains of bacteria was far from conclusive.

Both the Swann committee and a U.S. Food and Drug Administration task force reporting in 1972 made a similar recommendation: Antibiotics for addition to animal feeds should be restricted to those which have little or no application for disease treatment in either humans or animals. In other words, if drugs are to be used for growth promotion, they should not overlap with those used for therapeutic veterinary and medical uses, in order to reduce the pool of drug-resistant bacteria.

Since the release of the Swann report, Britain has banned the addition of penicillin and tetracycline to animal feeds, restricting their use to veterinary prescriptions. This move gained the support of the World Health Organization. There are, at present, similar moves afoot to do likewise in the United States.

Some European countries have gone a step further, limiting the veterinarian's access to antibiotics as well. There are reports indicating that in countries such as Germany and Holland, where the ban on tetracycline has been rigorously enforced in this manner, there has been a significant decrease in resistant populations of *Salmonella* bacteria.

In July 1977, Health and Welfare Canada announced a policy to reduce the unnecessary use of antibiotics in animal feeds:

"Some of the claims for specific antibiotics used in the mass medication of livestock for prevention of disease may not be valid in the light of present scientific knowledge. Furthermore, some of the antibiotics presently being used in Canada may no longer be efficacious for stimulation of growth or improved feed utilization, due to bacterial resistance.

"The Health Protection Branch intends to reduce unnecessary use of antibiotics for growth promotion and disease prevention by:

(a) continuing the present policy of permitting new antibiotics useful in human therapy to be sold only on prescription when used on animals;

(b) requiring manufacturers to produce new supporting data on the efficacy of all antibiotic growth promotants used in Canada and banning those shown not to be effective;

(c) requiring manufacturers to produce new supporting data for selected antibiotic uses in feeds where there are doubts regarding their efficacy in disease prevention. Uses which are not justified by these new data will be prohibited.

"It is not expected that this action will increase the cost of food to consumers.

"Based on available evidence, a total ban on the use of antibiotics for disease prevention and growth promotion in animals is unwarranted. These substances are essential for the production at reasonable costs of animal foods under the large scale, intensive conditions utilized by Canadian farmers."

This federal policy is a step in the right direction. Note, however, that it does nothing to reduce the use of antibiotics, such as penicillin, tetracycline, and streptomycin, that are already permitted in feeds and other non-veterinary (that is, farmer-administered) applications—the drugs for which the bacterial resistance problem is greatest.

Further protection against drug residues in animal products is warranted. A 1979 report to the Ontario Ministers of Agriculture and Food and of Health from a committee of physicians and veterinarians made, among its recommendations, the following suggestions:

● Feed and drug labels should be revised so that warnings, directions for use, and withdrawal periods are more clearly visible.

● A more intensive monitoring program for drug residues should be established at packing plants.

● Penalties should be imposed on anyone found guilty of misusing drugs and thereby causing harmful residues in animal products.

Heated debate surrounds the predicted economic effects of reduced livestock drug use. Organizations contributing comments for the 1979 Ontario report (above), including cattle, poultry, and pork producers' groups, had this to say: "At this time, removal of selected antibiotics from use in animal agriculture would likely result in an initial increase in food cost, possibly followed by some recovery, difficulty in controlling disease with a larger number of condemnations and marketing of lower quality animal products, possible marked changes in production procedures and management techniques and an unpredictable response by producers to these changing demands."

The U.S. Congress Office of Technology Assessment's 1979 study similarly concluded that the economic consequences of banning these drugs could be significant over the short term, assuming that no safe alternative drugs were found to replace those removed. The greatest impact would be on poultry production, the least on beef. Among the four key groups of antibiotics—penicillins, tetracyclines, sulfonamides, and nitrofurans—the greatest impact relative to total animal production would be through banning tetracycline, and pork and chicken production would be hit hardest.

But the U.S. report notes that "the long-term consequences are less certain, probably resulting in small decreases or no changes in production and small increases in both consumer prices and overall producer incomes."

Some alternative medications and growth promotants are already available. For example, Rumensin, a growth-promoting but non-hormonal and non-antibiotic feed additive approved for use in Canadian cattle production in 1977, looks promising. By altering the acid balance in the cattle's rumen, the chemical increases feed efficiency; it leaves no tissue residues.

There are probably as many testimonials from animal producers who have practically eliminated drug use without significant loss of income as there are from crop farmers who have "gone organic" and remained in the black. The common thread seems to be attention to detail—monitoring the crop or herd in order to prevent pest or microbial outbreaks and nipping them in the bud by non-chemical means when they do occur.

Individuals with severe drug allergies cannot wait 5 or 10 years for medicated meat to become a thing of the past. Some are resorting to eating only the meat of wild game, or seeking out the few producers of undrugged beef, "free-range" eggs, and the like.

What You Can Do

• Encourage governments to restrict the use of antibiotics and hormones in livestock production to therapeutic (disease-treating) uses only.

• If you have severe reactions to some drugs, seek out wild game, and/or meat and dairy products from undrugged sources. (This is no mean feat; contact the Human Ecology Foundation—see Appendix D.)

8 Toxic Moulds in Natural Foods

NINETEEN SEVENTY-TWO and 1973 were bad years in the midwest United States Corn Belt. Unusually wet weather delayed planting in the spring of '72 and harvesting that fall. The energy crisis meant shortages of natural gas, limiting the amount of corn that could be dried artificially after harvest. Some of that year's crop was not harvested till January of 1973. By the next summer, Food and Drug Administration director Sherwin Gardner reported that difficulties in harvesting and storing the crop had become "a national problem." The reason: Wet weather had allowed the grain to become contaminated with certain moulds which produce chemicals toxic to animals and humans. Toxins isolated from the 1972 corn harvest included aflatoxin B_1 (a carcinogen), zearalenone (a hormonal substance) and T-2 tricothecene toxin (which has a wide variety of acute effects, including heart failure when the toxin is ingested at high doses).

This chapter deals with mycotoxins, which are naturally occurring poisons produced by certain types of moulds (fungi) that may infect food crops. Particular attention will be paid to aflatoxin and its presence in peanuts and peanut butter. Methods will be outlined to reduce food contamination by aflatoxin and other mycotoxins such as ergot.

Mycotoxins

Until the early 1950s, moulds (fungi) were looked upon either merely as nuisances or in some cases as producers of beneficial substances. Moulds on grains, fruits, and vegetables detract from their appearance and cause some degree of economic loss due to blights, rusts, and spoilage. Other fungi are useful, though, in making blue, Camembert, and Roquefort cheeses. The most well known beneficial mould has been a species of *Penicillium,* producer of medicine's pet antibiotic. But if fungi can produce substances such as penicillin that are toxic to bacteria, it seemed reasonable to suspect other fungi of being able to produce substances toxic to animals and humans. Advances in mycology (the study of moulds) over the last 40 years have proven this to be the case. Dotted all over the globe and going back centuries, mysterious outbreaks of disease, seemingly related to consumption of mould-contaminated grains, were finally explained.

Mycotoxins are chemicals that appear to have no function (or toxicity, of course) in the moulds which produce them but which are toxic in varying degrees to birds and mammals. The five genera (groups of species) sporting the most mycotoxins are *Aspergillus, Fusarium, Alternaria, Cladosporium* and (yes!) *Penicillium. Table 25* lists some of these toxins and their effects at significant dosages. Grains and nuts are the foods most often contaminated with toxic moulds, fruits and vegetables less so. Eating food that contains only traces of mycotoxins seldom produces immediate or dramatic reac-

---

Table 25 / *Health Effects of Some Mycotoxins*

Toxin	*Effect*
aflatoxin B_1	hepatocarcinogenic (causes liver cancer)
citreoviridin	neurotoxic (toxic to the nervous system)
ergot alkaloids	convulsive, gangrenous, hallucinogenic
ochratoxin A	nephrotoxic (toxic to kidney) and teratogenic
patulin	respiratory failure
rubratoxin	teratogenic
sterigmatocystin	hepatotoxic (toxic to liver)
T-2 trichothecene toxin	emetic (causes vomiting)
zearalenone	estrogenic (having reproductive/ hormonal effects)

tions. One may feel chronic, vague ill health but the connection between such effects and earlier consumption of contaminated food can be easily overlooked. Furthermore, food harbouring mycotoxins does not often appear or taste "mouldy," so negligence on the part of the consumer cannot always be to blame when sickness occurs.

Mycotoxin contamination of food crops is sporadic, depending mainly on the weather in any one growing season. Adequate moisture is the critical factor determining fungal invasion of a crop; temperature is important, but to a lesser extent. Countries with hot, humid climates suffer most from mycotoxin contamination.

Aflatoxin and the Perils of Peanuts

In 1960, over 100,000 young turkeys in England died from an ailment dubbed "turkey X disease." The birds had dined on peanut meal contaminated with aflatoxin and had succumbed to irreparable liver damage. This episode ushered in the "aflatoxin era" and heightened concern about mycotoxins in general.

Aflatoxin derives its name from the mould species that most commonly produces it, *Aspergillus flavus.* There are several forms of the toxin, the most abundant being aflatoxin B_1. Given the proper conditions of moisture and temperature, aflatoxin can contaminate virtually any grain, fruit, or vegetable. The list includes peanuts, Brazil nuts, corn, wheat, soybeans, cottonseed, rice, spices, peppers, peas, and yams. Peanuts seem to be especially vulnerable.

Development of the mould begins in the field, right after harvest and before the nuts have been dried to a moisture level that no longer allows fungal growth. Broken and otherwise damaged (by insects, for example) shells and kernels are more prone to *A. flavus* attack than unblemished, intact ones. Once contaminated in the field, the nuts can suffer further mould damage through improper (that is, damp) storage. Once infected, peanuts cannot be rid of aflatoxin through further drying or application of fungicide.

People can be exposed to aflatoxin either directly, by eating tainted crops, or indirectly through the edible products of animals given tainted feed. Whole peanuts and peanut butter are often contaminated, whereas peanut oil rarely is, since the aflatoxin is retained almost entirely in the meal or cake remaining after the oil is pressed out of the nuts. Refining of the crude oil by treatment with alkali and bleaching agents reduces any residual aflatoxin to 1 part per billion (ppb) or less.

The brunt of the peanut problem, in terms of acute toxicity, is borne by livestock fed aflatoxin-laden meal as part of a compounded ration. The peanut meal which caused England's "turkey X disease" contained 7,000 to 10,000 ppb aflatoxin. Feed companies and livestock producers now generally recognize that aflatoxin-laden peanut meal should not be compounded indiscriminately into animal feed rations, not only due to factors of acute toxicity, but because of the economic impact of reduced growth and productivity of livestock given feed containing aflatoxin.

Aflatoxin contamination of animal products for human consumption is minimal, and appears to be restricted to milk and milk products in geographical areas where tainted feed occurs sporadically. The chemical is not readily stored in meat or eggs. Only about 1 percent of ingested aflatoxin is excreted in cow's milk. However, it resists destruction by pasteurization or drying of the milk, and binds tightly to the milk's casein (protein) fraction, indicating that casein-bearing products such as cheese may contain magnified levels of aflatoxin.

Mycotoxin research has focussed on aflatoxin because it is one of the most potent known carcinogens. It is *the* most active promoter of liver cancer, producing tumours in laboratory animals at the lowest analytically detectable level of 1 ppb. As is the case with many environmental sources of chronic disease, it is difficult to establish a cause-and-effect relationship between exposure and human cancer. However, there is some indication from population studies in southeast Asia and southern Africa (particularly in Mozambique) that liver cancer is significantly higher among groups who consume diets high in aflatoxins.

A risk assessment performed by the U.S. Food and Drug Administration concluded that as many as 66 cancers per 100,000 persons could develop from normal or expected ingestion of aflatoxin-contaminated peanut and corn products of the United States. The chemical also causes birth defects (i.e., it is teratogenic) and alterations in gene structure (mutations).

Although data are scanty, problems with aflatoxin in Canadian crops appear to be rare, since our climate is not hot and humid enough for *Aspergillus flavus* to thrive. Even domestic peanuts, introduced as a commercial crop for the first time in Ontario in 1980, have shown no sign of harbouring aflatoxin in a decade of field trials. Most of the peanuts and peanut products consumed in Canada are imported from the United States, and aflatoxin has been found in samples from most peanut-growing states.

Half of the peanuts we eat are in the form that sticks to the roof of your mouth—peanut butter. In 1978, the Consumers' Union, publishers of *Consumer Reports* magazine, found aflatoxin in at least one jar of every brand and type (smooth, crunchy) of peanut butter tested. Eighty-eight percent of the samples contained detectable aflatoxin, three exceeding 20 ppb, the U.S. legal tolerance. But just to demonstrate how seasonal aflatoxin contamination is, the American group noted that similar testing in 1972 detected aflatoxin in only 18 percent of the samples (all below 5 ppb) and in 1973, in only 11 percent of the samples (all below 10 ppb). No single brand, whether "supermarket-style" or "natural" (no added ingredients) came through consistently free of aflatoxin.

Total intake of aflatoxin in the Canadian diet is lower than that found in most countries, primarily due to our cold climate. Imported peanuts and peanut butter are of concern because of the quantities of these cheap, nutritious, "fun" foods that we eat. Recent research by Health and Welfare Canada indicates an average aflatoxin level in peanut butter marketed in Canada of about 4 ppb. Toxicologists in the health department are zeroing in on aflatoxin in the diet of Canadian children, because of their high intake of peanut butter relative to their weight.

Aflatoxin-containing peanut meal and corn fed to dairy cattle have in the past created problems in milk and milk products in some parts of the United States, especially the warmer, southern states. The American practice, now, is to divert any aflatoxin-contaminated feed away from dairy cattle and toward beef cattle. Canadian grains, however, are not particularly susceptible to contamination and peanut meal is not imported into Canada for use in livestock feed, so domestic dairy products are not a significant source of aflatoxin.

Aflatoxin is the only mycotoxin with a tolerance set down in Canada's *Food and Drugs Regulations*. No nut or nut product sold in Canada may legally contain more than 15 ppb aflatoxin. Each lot of raw peanuts brought in from south of the border is accompanied by a certificate of analysis from the U.S. Department of Agriculture. The peanuts may contain up to 25 ppb aflatoxin, which must be reduced to the 15 ppb limit through processing. Roasting destroys up to half of the aflatoxin. Mechanical and manual sorting to remove damaged or discoloured kernels further reduces toxin levels. Rejects from the sorting process are diverted to oil production or animal feed if the contamination is minor and to non-food/feed uses, such as fertilizer, if the contamination is serious.

Peanut processors are responsible for ensuring that aflatoxin in peanuts and peanut butter sold to the consumer does not exceed the legal limit. Health Protection Branch inspectors monitor processing plants on a regular basis, checking aflatoxin levels and recommending measures to further reduce these levels when necessary. Product recalls due to excessive aflatoxin residue are possible though uncommon (see *Table 26*).

Table 26 / *Aflatoxin-Related Enforcement Activities by the Health Protection Branch, Health and Welfare Canada, October 1978–March 1980*

Product	No. of Recalls	Amounts (approx.)	Reason for Recall
Peanuts, in shell	1	4,266 lb.	Contained aflatoxin in excess of 15 ppb (exact figures not available)
Peanuts, shelled	1	90 lb.	
Peanut butter	2	1,040 lb.	
Brazil nuts	1	4,355 lb.	
Brazil nut chocolate bars	1	15,708 boxes	
Pistachios	4	160,294 lb.	

Note: There were no convictions during this period.

Source: Health and Welfare Canada, Health Protection Branch. Unpublished data, May, 1980.

Clearly, *no* food should contain *any* cancer-causing aflatoxin. The question mark is the extent to which aflatoxin contamination of nuts and grains is avoidable. Since there's not much that can be done about the weather, emphasis in aflatoxin reduction schemes has been on proper drying of the crop—be it peanuts, corn, or wheat—at the farm level and proper sorting and storage at the processing level.

Ironically, the armour of modern agriculture has some chinks in it that compromise programs aimed at the eradication of aflatoxin (indeed, all mycotoxins). Herbicides are thought to lower the resistance of plants to invasion by mycotoxin-producing moulds. Also, large, highly automated harvesting equipment require that the crop be harvested a bit on the wet side, which presents the farmer with a drying problem. This equipment also increases physical (mechanical) damage to the crop, opening the door to more widespread fungal attack.

In addition to improvements in weed control and harvesting techniques, and more crop rotation to prevent carry-over of fungal infection of the soil from year to year, a further initiative in the battle against aflatoxin is the attempt to breed strains of crop species that show better resistance to mycotoxin attack. Mould contamination begins in the field and that's where it should end. If, on the farm, it can be prevented from starting at all, then it does not have to be controlled further along the production line.

Health and Welfare Canada would like to lower its aflatoxin tolerance in foods below the current 15 ppb. The legal limit should be the residue which is absolutely unavoidable in accordance with the best known agricultural and manufacturing practices.

Since aflatoxin contamination is not obvious to the eye or palate, the consumer is at the mercy of the peanut production system and its surveillance of the quality of its products.

Other Mycotoxins of Concern

Aflatoxin is public enemy number one when it comes to exposure to mycotoxins in the Canadian diet, and especially so because of its carcinogenic potential. But while aflatoxin becomes a Canadian problem almost exclusively through the importation of contaminated foods from warmer climes, there are a few, less frequently encountered mycotoxins for which the problem starts right here at home.

● Ergot

In 1978, the incidence of ergot in Canadian western red spring wheat was abnormally high.

Assisted by a cool, wet spring, the fungus *Claviceps purpurea* produces visible infections called "ergot bodies" on several cereal grains, especially rye and triticale (a wheat/rye cross), but also wheat, barley, and corn. Alkaloid chemicals produced by the ergot bodies have powerful pharmacological effects. (These chemicals are related to the hallucinogenic chemical LSD.) At high levels, ergot can cause convulsions and abortion. Smaller amounts consumed over an extended period lead to swelling and inflammation, burning and freezing sensations ("St. Anthony's Fire"), and restricted circulation, bringing on gangrene.

Unlike aflatoxin, ergot has long been recognized as a mycotoxin. Contaminated rye was reported in ancient Greece and ergot epidemics ravaged Europe from the tenth century up to modern times. Disease and death may befall both livestock consuming tainted feed and people eating bread from tainted flour. There has not been a recorded outbreak of ergotism in humans since 1951. However, livestock still occasionally show symptoms of ergot poisoning and lowered productivity of farm animals has sometimes been traced to small amounts of ergot in the feed. Canadian farmers are required to ensure that no animal feed contains more than 0.1 percent ergot.

Grain grading standards in most countries are strict enough to prevent grain contaminated with ergot above a certain small percentage from reaching commercial food channels. "Ergoty" grain is easily identified by the hard, black ergot bodies, about 80 percent of which can be removed from a contaminated lot with grain-cleaning equipment, although the procedure is expensive and time-consuming. While the ergot bodies may go unnoticed once the grain is ground into flour, this processing, as well as the time in storage, causes a great proportion of the toxic alkaloids in the ergot bodies to break down to harmless by-products.

As Klaus Lorenz, food scientist at Colorado State University, puts it: "Only ignorance, carelessness, or gross negligence on the part of a miller in a small community could cause another outbreak of ergotism in modern times. There always have been and always will be ergot infections and a possible danger to human health, but man has learned to minimize the effect and keep any contamination below the maximum levels permitted."

Regulations under the *Canada Grain Act* specify permitted levels of ergot in various grains. For example, Canadian western red spring wheat may contain from 3 ergot bodies per 500 grams (no. 1 grade) to 24 bodies per 500 grams (no. 3 grade). Levels permitted in rye are higher and are measured in percentages. No. 1 Canadian western rye may contain 0.05 percent ergot and no. 3, 0.33 percent. Any rye exceeding the 0.33 percent level of contamination is classed as "ergoty." There is no tolerance given in the *Food and Drugs Regulations* for ergot in wheat or rye flour, although a level of 0.05 percent appears to be acceptable in most countries.

As with aflatoxin, control of ergot requires good farm management. Agriculture Canada research scientist Christopher Young says that "ergot cannot be controlled by treatment with chemicals, and resistant varieties of cereals and grasses are not yet available. To control the disease, one should use ergot-free seed, apply crop rotation, and follow cultural practices that will reduce the incidence of ergot in grasses and its spread to cultivated crops."

● Zearalenone

Zearalenone is unique among the mycotoxins because it has a commercial use. Zearalenol, another form of zearalenone, is marketed in a very low dose under the name Ralgro, as a growth-promoting ear implant for cattle and sheep. Its effects derive from the fact that it is a hormonal compound, related in action to estrogen. (See Chapter 7 for further discussion of hormones used in rearing livestock.)

But what is good for cattle turns out to be not so good for swine. Corn, the main ingredient in swine feed in Canada, is prone to invasion by the fungus *Fusarium roseum,* producer of zearalenone. Swine fed contaminated corn experience infertility at levels as low as 100 parts per million. No livestock producer can afford infertility on a regular basis, and farm management practices aimed at minimizing zearalenone contamination are in order. But, as with other mycotoxins, weather conditions are a key determinant of zearalenone production, so its incidence varies from year to year.

At present, it is unclear whether zearalenone is a significant human health hazard, either through direct consumption of contaminated corn or by eating products from corn-fed animals. But this mycotoxin is suspected of being weakly carcinogenic. Furthermore, it often occurs together with other mycotoxins such as aflatoxin; possible synergism between the toxins needs more investigation.

● Pink Mould

In the summer of 1980, wet and humid weather caused part of the Ontario and Quebec winter wheat crops to be infected with a *Fusarium* fungus that was referred to as "pink mould." The toxin produced by the fungus causes vomiting, weight loss, and has effects on disease immunity in cattle, but its effects on humans are unknown.

Early in September 1980, the federal Health Protection Branch warned baby food manufacturers not to use eastern winter wheat in their products until more toxicological information on the mould is available. The toxin, believed to be of the tricothecene type, was present at levels of 0.01 to 6 parts per million in the wheat.

Later in the same month, Health and Welfare ruled that none of the 1980 Quebec winter wheat crop would be used for human consumption. (The Ontario crop was not as badly infected with the mycotoxin.) The federal health department stated that affected wheat crops could be used for animal feed, as long as Quebec farmers used it at a level no greater than 10 percent in the feed ration. The limit for the Ontario crop was set at 25 percent.

Growers of *Fusarium*-infected wheat claimed that the fungus threat had been poorly handled by the federal government, especially the Canadian Grain Commission (which temporarily froze exports of mouldy Ontario winter wheat) and was exaggerated in the media.

When more is known about this mycotoxin, a better name than "pink mould" should be coined, since other moulds, such as zearalenone, are also pink.

No one can afford mycotoxin-contaminated food products. Multi-million-dollar international grain deals can fall through if the grain is tainted with aflatoxin, ergot, T-2 tricothecene toxin, zearalenone, or other mycotoxins. Every link in the food production chain can be hurt—farmers, handlers, retailers, and consumers. Constant vigilance is the all-important weapon in the battle against toxic moulds.

What You Can Do

- Store peanuts, indeed all nuts, grains, seeds, and beans, in sealed containers in a cool, dry place.

- Store peanut butter in a sealed container either in the kitchen cupboard or in the refrigerator.

- Never eat any shrivelled, discoloured, visibly mouldy, or off-taste nuts of any kind.

● As the Health Protection Branch advises, "When in doubt, throw it out." This warning applies to all mouldy foods, those obviously so as well as those suspected of harbouring invisible mycotoxins.

9 Organic Pollutants (Alphabet Soup)

- **Billings, Montana, July 1979**

At the meat packing plant of the Pierce Packing Company, about 200 gallons of electrical transformer oil containing polychlorinated biphenyls (PCBs) leaked from a transformer in a storage shed into the plant's wastewater system. (The water is dredged for meat by-products to be used in animal feed rations.) The repercussions of the spill, in the form of contaminated feed, spread to 17 states and into the province of British Columbia. About 18 million tainted eggs were sold and consumed before the accident was uncovered in late August. Following the pathways taken by the toxic feed eventually led to the confiscation of some 400,000 pounds of poultry and 16,000 pounds of pork. Low levels of PCBs, linked to cancer as well as effects on the liver, nerves, and the disease immune system, were found in chicken, turkey, and eggs in B.C. as a result of the use of PCB-contaminated tallow, imported from the Montana plant, at three Abbotsford, B.C. feed mills.

- **St. Louis, Michigan, Spring 1973**

At the Michigan Chemical Company (now Velsicol Chemical Company), two products being manufactured simultaneously were NutriMaster, a cattle feed supplement containing magnesium oxide, and FireMaster, a highly toxic fire retardant chemical containing polybrominated biphenyls (PBBs). A labelling error led to the shipping of FireMaster to Farm Bureau Services Inc. in Battle Creek, Michigan, which mixes and markets livestock feed.

By the fall of 1973, many farm families in the state were struck with bizarre illnesses such as enlarged livers and spleens, rashes, stiff (even squeaking) joints, fatigue, loss of coordination, memory lapses, and inability to do simple tasks and make routine calculations. Their livestock didn't think much of PBBs either. Thousands became sick and died.

When the source of the contamination was finally located about a year later, 800 farms were quarantined; 33,000 cattle, 6,000 swine, 1,300 sheep and 1.5 million poultry were confiscated. The PBB accident has been called the worst case of mass poisoning in U.S. history. Eight million of Michigan's 9.1 million residents are carrying PBBs around in their bodies, including babies fed PBB-tainted breast milk. The PBBs found their way into Canada in beef imported from Michigan.

● **Ottawa, Ontario, March 1977**

Thirteen dairy farmers in the Ottawa River valley advised their feed supplier, Ritchie Feed and Seed Ltd., that their cattle were refusing to eat their feed. Agriculture Canada inspectors traced the problem to a Canadian National Railways boxcar that had been carrying a load of the wood preservative pentachlorophenol (PCP) to Alberta. The boxcar was not properly cleaned out and returned to Ontario with a load of PCP-contaminated feed oats. The foremost fear was that milk would become poisoned with dioxins, a group of highly poisonous chemicals that are formed during the production of PCP, possibly including TCDD (tetrachlorodibenzodioxin), probably *the* most toxic chemical known to man. Fortunately, government testing of the suspect milk did not uncover any dioxins, although measuring equipment at the time was not particularly sensitive to minute amounts.

This chapter will examine how pollution of the air and water, as well as other improper handling of toxic chemicals, leads to contamination of food. It will then zero in on organic (carbon-based) pollutants. Much of the chapter will be devoted to a discussion of one group of organic chemicals, the polychlorinated biphenyls (PCBs). Their presence in fish and the problems posed by PCB-contaminated mothers' milk will be highlighted. Other organic pollutants to be looked at are dioxin, mirex, benzene, and polybrominated biphenyls (PBBs).

"Everything Leaks"

When pesticides are applied directly to food crops, it is to be expected that residues of these toxic chemicals might appear in food products. After all, treating food to a dose of DDT, malathion, or carbofuran is deliberate and planned. But treating eggs to a dose of PCBs or PBBs as a result of tainted feed and treating fish to a dose of mercury from polluted water is environmental contamination of food in its most insidious form—accidental, incidental lacing of the food supply with the by-products of industrialization.

Robert Paehlke, a professor of political studies at Trent University in Peterborough, Ontario, has summed up the problem in two words: "Everything leaks." Since we human beings are fallible and the machinery and processes of our industrialized society imperfect, it is almost inevitable that some of the chemicals we intend to contain in one place are going to end up inadvertently in another.

This is *not* to say that further precautions cannot be taken to reduce the chances of toxic leaks. Indeed they can. But first we must recognize the nature of the shortcomings in the ways we currently handle toxic chemicals.

Says Paehlke: "We must simply accept that any substance used in large quantities will one way or another, sooner or later, be released in some amount into the environment. So-called 'contained' uses are a myth. The 'sealed' condensers in fluorescent light fixtures (which contain PCBs) ultimately go to the dump, rust, and spill their contents. Production facilities age, warehouses burn down, by-products and waste products are never perfectly handled, all known modes of transportation have unanticipated accidents, older workers become habituated, newer workers are imperfectly trained, and many consumers pay no attention to instructions regarding use and disposal. Small amounts are released—and in nature, remarkably dilute amounts of some substances can return to haunt us in unexpected ways."

Few discoveries of the sources of pollutants in food are as clear cut as the examples that opened this chapter. Although in all three cases it took some months to uncover the sources of contamination, those sources were ultimately found. This type of pollution is often called "point-source" pollution, in that the toxins emanate from one distinct location. Perhaps the most notorious example of ongoing point-source pollution of food in Canada is the tainting of fish in the Wabigoon and English Rivers by mercury from the Reed Paper mill in Dryden, Ontario.

Point-source pollution is no less harmful than diffuse or non-point-source pollution; it's just easier to remedy and those prosecuting the offenders have a greater chance of success because the offenders are readily identifiable. When PBBs ended up in Michigan meat and the offending feed and chemical companies were identified, lawsuits flew thick and fast.

On the other hand, when DDT shows up in fish caught in our lakes and rivers and when lead appears in vegetables grown near busy highways, it's virtually impossible to pinpoint and take direct action against each culprit, simply because there are so many of them. Non-point-source pollution is, in effect, a great number of small point sources, from the thousands of farmers who used DDT 10 or 20 years ago and who contaminated waterways through run-off from their farms, to the millions of automobile owners still fueling their cars with leaded gasoline.

Contamination of food by pollutants can be remarkably indirect. For example, air pollution can lead to water pollution which, through fish, can lead to food pollution. The International Joint Commission, a Canada/United States agency which performs research and makes recommendations on pollution in waterways shared by the two countries, estimates that air pollutants, or "atmospheric loading," may be responsible for up to 40 percent of the loadings of some pollutants to the Great Lakes, including organic chemicals like PCBs and metals such as lead.

As pointed out by Joe Castrilli and Elizabeth Block of the Canadian Environmental Law Association, environmental legislation in Canada is, in theory, broad enough to prohibit not only point-source but also non-point-source pollution. In practice, however, the latter escapes relatively unscathed because the laws "cannot be used when the violators cannot be identified, or when there are so many that it is impossible to go after them all. Blanket prohibitions, but piecemeal enforcement, will not change business as usual."

The solution to food pollution, then, is not more court cases but, rather, what is usually called "reduction at source," either voluntarily or by new legislation. Well-known examples include the replacement of lead in gasoline by less harmful antiknock chemicals, and the substitution of PCBs in electrical equipment by other insulators. (It is questionable, though, whether ethylene dibromide, a replacement for lead in gasoline, is a step up. Engine redesign to permit the burning of lower-octane fuel would be a better solution to the lead problem.)

Since the pollutants of greatest concern are persistent ones, those that decompose very slowly, it follows that even if today we shut off all the pipes, park all the cars and trucks, and ban all the DDT, the problems of food pollutants will remain with us for years to come. In this situation, the best we can do is become knowledgeable about which foods are likely to be most contaminated and in what localities…and avoid them. Residents of industrial areas, for example, should be warned about eating produce from backyard gardens or fish from local lakes and rivers.

Organic Pollutants (Alphabet Soup)

Pollutants in food can be divided into two categories, organic and inorganic chemicals. Organic chemicals must not be confused with the term "organic" as it is used to describe ecological agriculture. (See Chapter 6 and the Glossary, Appendix B.)

Most of the organic chemicals which show up as food pollutants belong to a group called halogenated hydrocarbons, a cumbersome but useful way to describe chemicals which contain carbon, hydrogen, and the halogen elements chlorine or bromine. (Organochlorine insecticides such as DDT are also halogenated hydrocarbons.) Since these organic pollutants have almost impossibly long chemical names, they are usually abbreviated to form a less than tasty bowl of alphabet soup—PCBs, PBBs, PCP, TCDD, and so on. Their chemical structures are similar to DDT, so it is no accident that they share many of the insecticide's unpleasant attributes—persistence in the environment, effects on the human nervous system, and definite links to cancer and birth defects.

These pollutants may even be affecting male fertility. The Florida State University research by Ralph Dougherty, mentioned in Chapter 5, found not only insecticides in the semen samples, but also PCBs and PCP. However, the link between fertility decline and pollutants is only conjectural at present.

A telling indication that organic pollutants may have significant human health effects is the link between pollutant levels in waterways and diseases in fish. R.D. Moccia and his colleagues at the University of Guelph in Ontario have hypothesized a link between organic chemicals in the Great Lakes and incidence of thyroid goiters in coho salmon in those lakes.

The common organic pollutants are discussed below. (See Chapter 10 for a discussion of metals, the most common inorganic pollutants.)

Polychlorinated Biphenyls in Food Other Than Fish

PCBs, of which there are over 100 different types in commercial use, were first manufactured in 1929. They used to be included in products as diverse as plasticizers for synthetic resins and carbonless copy paper. But their use is now restricted to the electrical industry where, because of their stability and fire-retarding properties, they still serve as a dielectric fluid for capacitors and transformers. (However, they are gradually being phased out and replaced with more environmentally acceptable chemicals.)

The two properties of PCBs that make them so attractive to the electrical industry—their stability over time and over a wide temperature range, and their solubility in other hydrocarbons—are the very properties that make PCBs a serious environmental problem. First, they are even more persistent than DDT, and second, solubility in hydrocarbons makes them soluble in the fats found in living organisms and so makes them subject to bio-concentration (see Chapter 5).

PCBs were first recognized as environmental contaminants in 1966 when Swedish scientists detected them in fish from the Baltic Sea and in eagle feathers. They may have been present before that time, but the analytical techniques in use until then are thought to have misinterpreted PCBs as part of the family of DDT residues under intense scrutiny in the 1960s. Therefore, some of the environmental and health impacts blamed on DDT may actually have been due to PCBs.

PCBs enter the environment through run-off from roads onto which waste oils containing PCBs have been sprayed as a dust suppressor, from the incineration of municipal wastes which may include carbonless copy paper, by leaking from landfill sites where PCB-containing equipment has been junked, from leaks in hydraulic and heating systems, from leaking transport trucks and train derailments, and from other accidents such as the December 1977 fire at an Ontario Hydro transformer station in downtown Toronto.

The suspected or proven human health effects of PCBs are legion.

Chloracne, a skin disorder, has been noted in industrial workers whose skin is exposed to PCBs and in the victims of "Yusho disease," caused by ingestion of PCB-contaminated rice oil in Japan in 1968. (It has been suggested, however, that the chloracne is caused by contaminants in the PCBs, such as furans, rather than by the PCBs themselves.) Other disorders suffered by Yusho victims, who numbered over 1000, included eye discharges, headaches, loss of appetite (anorexia), vomiting, fever, and visual disturbances. The rice oil contained PCBs at exceptionally high levels—2,000 to 3,000 parts per million.

At the lower levels typically found in contaminated foods—of the order of 0.1 to 10 ppm—PCBs have been found, in laboratory studies with animals, to induce cancers, liver disease, birth defects, and effects on disease immunity. PCBs are highly suspected but as yet unproven human carcinogens. They are also thought to interact synergistically with other chlorinated organic chemicals such as DDT, producing increased activity of liver proteins called hepatic microsomal enzymes. About 90 percent of any PCB that is eaten in food is retained in the body, concentrating in fat (adipose) tissue and breast milk.

Like the 1979 Montana episode and the 1968 Japan poisoning, point-source PCB contamination of food is unusual. But accidents do happen. Other accidents include: contamination in 1971 of 800,000 tons of animal feed (much of which got into meat and eggs before the problem was recognized), due to a leak in the heating system of a feed mill in North Carolina; leakage of a PCB-containing sealant in silos in Ohio, Georgia, and Florida in 1970, leading to contamination of silage fed to dairy cattle and PCB residues in their milk; and detection by the Campbell Soup Company of high levels of PCBs (about 27 ppm) in the fat of chickens in New York state in 1971 and voluntary rejection of the tainted poultry by Campbell's.

As a result of less easily identifiable, diffuse PCB pollution, low PCB residues have shown up in a number of foods. The U.S. Food and Drug Administration has detected PCBs in fish, cheese, milk, eggs, potatoes, oysters, crabmeat, candy, meat, poultry, and cereals. Data on the PCB content of meat and poultry in Canada in the early 1970s are shown in *Tables 27* and *28*. In all cases, PCB residues, though very low, increased between 1969 and 1974, and appear to be lower in the western and Atlantic provinces than in Ontario and Quebec.

PCBs were first detected in the "typical" Canadian diet in Health and Welfare Canada's 1976–78 total diet study (described in Chapter

5). Previous studies in 1969, 1970, 1971, 1972 and 1973 had uncovered no PCBs.

Table 27 / *PCBs in Canadian Meat and Poultry, 1969–1974*

Time Period	Average PCB Residues (parts per million)		
	Beef	*Pork*	*Poultry*
Oct. 1969–Apr. 1970	0.00018	0.00094	—
Apr. 1970–Oct. 1970	0.00025	0.00103	—
Oct. 1970–Apr. 1971	0.00033	0.00116	—
Apr. 1971–Apr. 1972	0.00048	0.00127	0.00312
Oct. 1971–Apr. 1972	0.00059	0.00168	0.00386
Apr. 1972–Oct. 1972	0.00081	0.00274	0.00335
Oct. 1972–Apr. 1973	0.00139	0.00463	0.00400
Apr. 1973–Oct. 1973	0.00214	0.00771	0.00421
Oct. 1973–Apr. 1974	0.00321	0.00902	—
Apr. 1974–Oct. 1974	0.00459	0.00967	0.00581

Source   P.W. SASCHENBRECKER (Agriculture Canada). "Levels of Terminal Pesticide Residues in Canadian Meat." *Canadian Veterinary Journal*, June, 1976, pp. 158–163.

Table 28 / *Regional Differences in PCB Residues in Canadian Meat and Poultry, 1969–1974*

Province	Average PCB Residues (parts per million)		
	Beef	*Pork*	*Poultry*
British Columbia	0.00012	—	0.00120
Alberta	0.00013	0.00361	—
Saskatchewan	0.00009	0.00126	0.00630
Manitoba	—	0.00008	—
Ontario	0.01747	0.00027	0.00630
Quebec	0.01231	0.00246	0.00300
Atlantic Provinces	0.00008	0.00028	0.00891
National Average	0.00800	0.00139	0.00410

Source: P.W. SASCHENBRECKER (Agriculture Canada). "Levels of Terminal Pesticide Residues in Canadian Meat." *Canadian Veterinary Journal*, June, 1976, pp. 158–163.

PCBs and DDT in Fish

Human exposure to PCBs in food other than fish pales in comparison to the levels experienced either in occupational settings or in a diet that relies heavily on fish from polluted waters. Since PCBs migrate to fatty tissues, those fish which are higher in fat and those which are predatory (eating smaller fish and bio-concentrating the PCBs) are the ones to watch. These include coho and chinook salmon, lake trout, carp, and catfish.

Low levels of PCBs and DDT (the latter from farm run-off) have been detected in fish from lakes and rivers all across Canada—see *Tables 29* and *30*. But the critical geographical area, as can be seen in

---

Table 29 / *Average Residues of PCBs Found in Canadian Fish*

Water Body	Fish Species	Average Residue (parts per million)
Isolated Areas		
Lake Minto (Northern Quebec)	lake trout	0.08*
Agricultural Areas		
Lake Simcoe (Southern Ontario)	rock bass, yellow perch, largemouth and small-mouth bass, white sucker, walleye, burbot, northern pike, eastern shiner	0.34
	lake trout	5.50
Creeks in Norfolk County (Southern Ontario)	rainbow trout	0.99
Prairie lakes	northern pike, sauger, wall-eye, yellow perch, black bullhead, common white-fish, white sucker, brook stickleback, shiners	<0.10*
Urbanized Areas (Great Lakes)		
Lake Superior	bloater, lake whitefish, lake trout, chub	{ 2.8* 0.6
Lake Huron	smelts, alewife, chub, kokanee salmon, splake, whitefish, walleye, coho salmon, carp, channel catfish, yellow perch	6.6* 3.9
Lake Michigan	chub	4.5
	bloater, white sucker, yellow perch	9.5*

* residues in the whole fish; other figures are residues in edible tissue

Lake Erie	smelt, white bass, yellow perch, chub, carp, catfish, walleye, freshwater drum, burbot, gizzard shad, alewife	4.2*
Lake Ontario	smelt, white bass, yellow perch, smallmouth bass, alewife, sculpin	15.4*
	yellow perch, white perch, rock bass	6.3*
Marine		
Bay of Fundy	herring, mackerel, plaice, white hake, ocean perch, cod	0.05

Source: National Research Council of Canada (Associate Committee on Scientific Criteria for Environmental Quality). *Polychlorinated Biphenyls: Biological Criteria for an Assessment of Their Effects on Environmental Quality.* NRCC Publication No. 16077. Ottawa: National Research Council of Canada, 1978.

Table 30 / *Average DDT/PCB Ratios in Canadian Dressed Fish*

Water Body	Date of Analysis	Average DDT/PCB
Lake Erie (Ontario shore)	1970–71	0.05
Lake Erie (Ohio shore)	1970–71	0.30
Lake Huron	1970–71	0.95
Lake Ontario	1970–71	0.23
Lake Michigan	1970–71	2.31
Lake Superior	1970–71	1.98
Muskoka lakes (Southern Ontario)	1970–71	1.88
New Brunswick inland waters	1970	1.10
Nova Scotia banks (marine)	1970	0.60
Lake Minto (Northern Quebec)	1970	1.46*
Lake Nipigon (Northern Ontario)	1969	4.57*

* residues in the whole fish; other figures are residues in edible tissue

Source: National Research Council of Canada (Associate Committee on Scientific Criteria for Environmental Quality). *Polychlorinated Biphenyls: Biological Criteria for an Assessment of Their Effects on Environmental Quality.* NRCC Publication No. 16077. Ottawa: National Research Council of Canada, 1978.

the tables, is the Great Lakes watershed and its outlet, the St. Lawrence River. This drainage basin, shared by Canada and the United States, is home to over 37 million people, their farms, and their industries. PCB and DDT residues in Great Lakes fish can be 10 to 100 times the levels observed in the edible portion of fish from less contaminated waters. Individuals in this area with a heavy diet of local fish, as well as young children and women of childbearing age eating *any* local fish, are considered the groups most at risk, and least protected by government-imposed maximum residue guidelines.

The Canadian government tolerance for DDT in fish is 5 ppm and the "guideline" (not in regulation) for PCBs in fish is 2 ppm. Research by scientists in the Ontario Ministry of Agriculture and Food and the Ministry of Natural Resources has indicated that while *average* (mean) residues are often below the tolerance or guideline, *individual* fish regularly exceed these levels—see *Table 31*. The toxic individuals are

Table 31 / *Some Differences Between Average and High Residues of PCBs and DDT in Great Lakes Fish (1968 to 1976) in Relation to Federal Guidelines**

Lake	Fish Species	Federal DDT Guideline	Average DDT (ppm)	High DDT (ppm)	Federal PCB Guideline	Average PCB (ppm)	High PCB (ppm)
Superior	lake trout	5 ppm	2.72	14.10	2 ppm	1.80	2.30
Huron	walleye	5 ppm	5.05	6.22	2 ppm	1.80	3.90
	lake trout	5 ppm	7.60	9.85	2 ppm	0.70	0.90
	bloater	5 ppm	5.18	6.43	2 ppm	5.20	6.40
Lake St. Clair[+]	longnose gar	5 ppm	1.19	2.35	2 ppm	1.50	4.00
	largemouth bass	5 ppm	0.41	0.88	2 ppm	1.30	4.30
Lake Erie	coho salmon	5 ppm	2.80	8.23	2 ppm	4.00	14.00
	smallmouth bass	5 ppm	0.17	0.28	2 ppm	5.80	9.30
	white bass	5 ppm	0.09	0.19	2 ppm	2.20	4.80

* Federal guidelines are for edible fish tissue
+ linking Lakes Huron and Erie

Sources: R. FRANK *et al.* "Residues of Organochlorine Insecticides and Polychlorinated Biphenyls in Fish from Lakes Huron and Superior, Canada—1968–76." *Pesticides Monitoring Journal,* September, 1978, pp. 60–68.

R. FRANK *et al.* "Residues of Organochlorine Insecticides and Polychlorinated Biphenyls in Fish from Lakes St. Clair and Erie, Canada—1968–76." *Pesticides Monitoring Journal,* September, 1978, pp. 69–80.

usually the larger, heavier, more mature fish which have accumulated the contaminants over a longer period of time.

Also bearing on the level of contamination is the site of catch. The Ontario government researchers found that in fish of the same species, DDT residues were 10 to 15 times higher in fish caught in streams emptying into the lakes than those caught in the open water of the lakes, where the pollutants have had a chance to become diluted.

Residues of DDT in Great Lakes fish have generally declined since the early 1970s in response to restrictions and eventually bans on its use. PCBs have shown no consistent trend, increasing over time in some species and decreasing in others.

A consequence of these findings of organochlorine residues in excess of "safe" levels has been the Ontario government's development of a Fish Contaminant Monitoring Program, an initiative which is unique in North America. Each year, the provincial authorities test fish from 200 to 250 lakes and issue three handbooks called *Guide to Eating Ontario Sport Fish*—one each for the Great Lakes, Southern Ontario and Northern Ontario.

As the handbooks put it: "Fishing is knowing about bait, tackle, and where the big ones are. Today it also means being informed about the amount and type of sport fish you can eat." The contaminants routinely checked for and reported in the handbooks are mercury (see Chapter 10), PCBs, DDT and the pesticide/fire retardant mirex (discussed later in this chapter).

The handbooks *are* handy. They list contaminant levels according to fish size; indicate where the fish are safe for non-pregnant adults but not for children and pregnant women; and state how often a fish meal from any one lake or river can be enjoyed. As well, notes the *Probe Post,* "the contamination problem provides a convenient excuse for the less proficient, who can now say they let the big one get away because a quick check of the Ontario Guide booklets showed the brute to be an undesirable meal."

However, a *Probe Post* critique of the 1980 guidebooks explained that only 878 of Ontario's 200,000-plus lakes and rivers have been tested so far. (In two waterways, the Wabigoon River and Clay Lake, *no* fish *at all* can be eaten, due to heavy mercury pollution from the Dryden pulp and paper mill.)

The guidebooks deal only with fishing for recreational purposes, providing a way for anglers to check the safety of their own catches.

Most of the commercially netted fish we eat is of marine origin and is less contaminated. The vast majority of the fish caught commercially in the Great Lakes is exported ("dumped") to countries such as Japan, France, and Germany, which have less restrictive standards for organic pollutants and mercury in fish. (The fish usually exceed our own federal government's guidelines for these toxins.)

PCBs in Mothers' Milk

When Ontario government research first indicated high levels of PCBs in fish in many lakes in the province, special warnings went out to expectant and nursing mothers and other women of childbearing age. The reason? In addition to concentrating in fat tissue as do other organochlorine compounds (see *Table 32*), PCBs find a "sink" in the fat of breast milk, preferring excretion in the milk to excretion in urine and feces. Japanese research has found that PCB levels in human breast milk are about 10 times higher than those found in human blood. Such information is of utmost importance simply because infants and children are much more susceptible to the toxic effects of PCBs.

Table 32 / *PCBs in Adipose (Fat) Tissues of Canadians*

(A) *National Data*

Average PCB residue in adipose tissue extracts: 0.907 parts per million (ppm)
Range of residues: 0.106–6.603 ppm
Percent of samples containing detectable residues: 100%

(B) *Regional Data*

Average PCB residues in adipose tissue extracts

Atlantic Provinces	0.727	ppm
Quebec	0.969	ppm
Ontario	1.070	ppm
Manitoba and Saskatchewan	0.499	ppm
Alberta and British Columbia	0.898	ppm

Source: JOS MES *et al.* "Polychlorinated Biphenyl and Organochlorine Pesticide Residues in Adipose Tissue of Canadians." *Bulletin of Environmental Contamination and Toxicology,* vol. 17, no. 2 (1977), pp. 196–203.

The infant's exposure to PCBs begins in the womb, since the chemicals can cross the placenta. PCB levels in breast milk are highest soon after delivery, declining steadily in the first eight months post partum. Also, the PCB levels in milk after a first delivery are higher than those observed after subsequent deliveries. Mothers who have been occupationally exposed to high levels of PCBs are usually advised not to nurse their infants at all.

Considerable work on PCBs in breast milk has been done by T. Yakushiji and his colleagues at the Osaka Prefectural Institute of Public Health in Japan. The research was motivated by occupational exposures in that country and by the accidental rice oil contamination in 1968 (the "Yusho disease" mentioned earlier). In a 1979 research paper they stated:

"The safety evaluation of PCB residues in human milk is very difficult. All conditions should be considered: concentration of PCBs, analysis of isomers [different kinds of PCBs] which may be variable in toxicity according to their difference of structures, the amounts of daily intake, the period of ingestion and the additive effects of organo-chlorine pesticides and other chemicals. Even if babies did not show any adverse clinical symptoms related to the toxic effects of chlorinated hydrocarbons, further studies concerning toxic effects of PCBs on human infants and understandings of long-term trends of chlorinated hydrocarbon levels in human milk are necessary."

Testing for PCBs in the milk of North American mothers was first carried out by the United States Environmental Protection Agency (EPA) in 1975. PCBs were found in 1,029 out of the 1,038 milk samples tested from 45 states, at an average level of 0.093 parts per million (93 parts per billion) in the whole milk. Of the 1,029 positive samples, 720 contained PCBs below the 0.05 ppm (50 ppb) detection limit; 70 had levels over 0.10 ppm (100 ppb), and 11 were over 0.20 ppm (200 ppb). The highest single reading was 0.960 ppm (960 ppb), from Alabama. EPA officials said they were uncertain whether these PCB levels posed health problems and did not advise the mothers to change their breast-feeding habits, but all women participating in the survey were notified of the results.

If such testing results are expressed in terms of the PCB concentration in only the *fat* of the milk, then the levels can be compared with PCB guidelines for other foods. A level of 2 ppm in the fat of human breast milk was not uncommon in the EPA survey. This residue is below the 2.5 ppm concentration which the U.S. Food and Drug Administration allows in the fat of cow's milk, but more than eight times the 0.2 ppm permitted in baby food.

Breast milk testing by the New York State Department of Health found PCB levels 5 to 10 times higher than the EPA averages in women who regularly consume fish from Lake Ontario. James Allen, a pathologist at the University of Wisconsin, issued this advice: "If I were a physician who had women under my surveillance who were living along the Great Lakes or the Hudson River area where they were eating fish four or five times a week, I would suggest they put their infants on formula."

A preliminary survey conducted across Canada in 1975–77 by the Health Protection Branch of Health and Welfare Canada found that PCBs in mothers' milk ranged between 1 and 68 parts per billion (micrograms per kilogram) in the whole milk, with an average of 12 ppb—see *Table 33*. The 12 ppb average was a substantial increase over the 6 ppb average found in 1970, although conclusions are hard to draw because of the limited amount of data. By contrast, other

---

Table 33 /   *PCBs in Mothers' Milk in Canada*

(A)   *National Data*

Average PCB residue in the whole milk: 12 parts per billion (ppb)
Range of residues: 1–68 ppb
Health and Welfare Canada maximum acceptable residue: 50 ppb*

(B)   *Regional Data*

Average PCB residues in whole milk

Atlantic Provinces	8 ppb
Quebec	10 ppb
Ontario	17 ppb
Manitoba and Saskatchewan	8 ppb
Alberta and British Columbia	15 ppb

Source: A.B. MORRISON. "Polychlorinated Biphenyls, Department of National Health and Welfare—Committee Report." *Information Letter (Health Protection Branch),* March 31, 1978.

Source: JOS MES and D.J. DAVIES. "Presence of Polychlorinated Biphenyl and Organochlorine Pesticide Residues and the Absence of Polychlorinated Terphenyls in Canadian Human Milk Samples." *Bulletin of Environmental Contamination and Toxicology,* vol. 21 (1979), pp. 381–387.

---

halogenated hydrocarbons in human breast milk, such as the insecticides DDT, dieldrin, and heptachlor epoxide, decreased over the same period.

In January 1978, Health and Welfare Minister Monique Bégin convened a committee comprised of members of the Canadian Paediatric Society, the Pharmacological Society of Canada, and toxicologists from both Canada and the U.S. to consider possible health problems related to PCBs in mothers' milk.

The committee had the following advice:

● Women from the Great Lakes basin and women who may have had occupational exposure to PCBs and who intend to nurse their babies should contact their physicians.

● Where PCB levels in the whole milk are over 50 parts per billion, the physician "should pay particular attention to the mother's history, the child's birth weight, general conditions and subsequent growth and development in advising the mother concerning breast feeding. In the light of currently available information, it was the Committee's opinion that in view of the benefits of breast-feeding in most instances, it would be advisable to continue nursing."

● Further research is required to determine if adverse effects occur in infants exposed to breast milk containing PCBs.

However, University of Guelph biochemist Stephen Safe is of the opinion that the Canadian government's acceptable level of 50 ppb for PCBs in human milk is at least five times too high. His research team found that some of the most toxic individual PCBs are retained by the body and can become concentrated in the milk.

"The PCBs in breast milk are five times more active than the commercial product," says Safe. "Because we're apparently preferentially bio-concentrating the more active and toxic PCBs, we may have been underestimating their potential harm."

Richard Frank, director of the Ontario government's Pesticide Residue Testing Laboratory in Guelph, notes that recent surveys of human milk in Ontario show 4 percent of samples to be above the 50 ppb "safe" limit set by Ottawa. Only 19 percent were below 10 ppb, the level Stephen Safe considers appropriate.

As mentioned in Chapter 5 with respect to DDT in breast milk, the benefit/risk assessment of human milk, a highly nutritious infant food often contaminated with pesticides and pollutants, is very difficult.

Coping with PCBs

In 1971, uses for PCBs were voluntarily restricted by the Monsanto Chemical Company, the only PCB manufacturer in North America, to so-called closed systems such as electrical equipment. In 1979, regulations developed under the federal *Environmental Contaminants Act* banned the use of PCBs in new products in Canada. Furthermore, existing capacitors and transformers that need refilling with new dielectric fluid must not be filled with PCBs; safer alternatives have been found.

The two remaining problems, then, are informing the public about past indiscretions which led to the PCB residues we're now finding in fish and human milk, and preventing future problems by proper disposal of existing PCBs. High temperature incineration appears to be the safest means of disposal (it breaks down the PCB molecule), much preferable to sanitary landfilling which allows the PCBs to leach into waterways. Other disposal mechanisms are currently being developed.

An Ontario provincial regulation still permits dust-suppressing road oil to contain up to 25 parts per million PCBs. (No other province regulates road oil composition.) It is expected, though, that Canada will soon follow the suggestion put forth by the United States Environmental Protection Agency that this tolerance be removed altogether. Oils for burning in such industrial establishments as cement kilns may contain up to 100 ppm PCBs in Canada. Neither country has banned the presence of PCBs in heating equipment of feed and

---

Table 34 / *Canadian Guidelines on PCBs in Food*

Food	*Guideline* (administrative maximum residue limit), parts per million
Milk and dairy products	0.2 (in the fat)
Poultry	0.5 (in the fat)
Eggs	0.1 (less shell)
Beef	0.2 (in the fat)
Fish and shellfish	2 (edible portion)

Source: Health and Welfare Canada, Health Protection Branch. Unpublished data, September, 1980.

---

food processing plants, although a limit of 50 ppm has been put forth in a proposed amendment to the U.S. *Toxic Substances Control Act.*

Canadian government guidelines (or "administrative maximum residue limits") for PCBs in foods are shown in *Table 34.* These are similar to the ones accepted in the United States, although that country also has guidelines for PCBs in animal feed, baby food and food packaging. (See Chapter 11 for a discussion of PCBs in food packages.) Considering the suspicion that PCBs are human carcinogens, we must question the appropriateness of guidelines allowing *any* level of PCBs in food.

Dioxins in Food

It's hard to imagine a substance which is 500 times more poisonous than strychnine and 10,000 times more poisonous than cyanide. It's equally hard to imagine that only a kilogram (about 2 pounds) of such a substance caused the evacuation of an entire town—Seveso, Italy—after an industrial accident. The chemical is 2,3,7,8-tetrachlorodi-benzo-p-dioxin, or TCDD for short. TCDD is the most toxic of a group of compounds called dioxins, but since it is the most notorious, talk of dioxin usually means talk of TCDD.

Dioxins, like PCBs, are halogenated hydrocarbons. They are formed as an unwanted contaminant in the production of a number of chemicals, most of which are pesticides—see *Table 35.* The type of dioxin formed is a function of the temperature at which the pesticide is being produced. The popular herbicide 2,4-D is produced at relatively low temperatures, and has been generally believed to be dioxin-free. However, in 1980, scientists at Agriculture Canada found dioxins in some samples of 2,4-D, though no TCDD was uncovered. Agriculture Minister Eugene Whelan stated late in the year that his department's findings might affect regulations on the use of 2,4-D.

By contrast, its cousins 2,4,5-T and 2,4,5-TP are produced at much higher temperatures which promote TCDD formation. Pentachloro-phenol (PCP) more often contains HCDD and OCDD than the more toxic TCDD.

Extremely small amounts of TCDD are toxic. TCDD at the level of 5 parts per *trillion* is a suspected carcinogen in laboratory animals. The no-effect level of HCDD for the production of birth defects in rats is about 1 part per *billion.* In man, overexposure to dioxin can cause

Table 35 / *Dioxins and Their Sources*

(A)   *Types of Common Dioxins* (total number of dioxins: 75)

tetrachlorodibenzodioxin (TCDD)*
hexachlorodibenzodioxin (HCDD)
heptachlorodibenzodioxin (hepta-CDD)
octachlorodibenzodioxin (OCDD)

(B)   *Products Containing Dioxins and Their Uses*

Product	Uses
2,4,5-T	Herbicide—rice crops, pasture land management, utility and highway rights-of-way, forest brush control
2,4,5-TP (Silvex, Fenoprop)	Herbicide—utility and highway rights-of-way, forest brush control, lawn products
	Growth Regulator—apple crops (to prevent premature dropping)
Pentachlorophenol (PCP or "penta")	Wood preservative
	Herbicide—roadside and driveway weeds
Trichlorophenol	Leather tanning
Hexachlorophene	Skin treatment (acne, burns, wounds), control of staphylococcal diseases, cosmetics preservative, disinfectant soaps

*The full chemical name is 2,3,7,8-tetrachlorodibenzo-p-dioxin

Sources: P. KRUUS and I.M. VALERIOTE (editors). *Controversial Chemicals: A Citizen's Guide.* Montreal, Quebec: Multiscience Publications Ltd., 1979.

Vermont Public Interest Research Group. "Dioxins in Vermont." *Not Man Apart* (Friends of the Earth), August–September, 1978, pp. 6–7.

chloracne, a skin disease caused by the body's attempt to get rid of the poison through the skin (as in the case with high exposures to PCBs). More subtle effects may include atrophy of the kidneys, cirrhosis of the liver, stomach ulceration, hemorrhaging, central nervous system damage, emphysema, and mental disturbances such as memory and concentration lapses, and depression.

Dow Chemical, a leading producer of chemicals containing dioxins as impurities, is on record as saying that "2,4,5-T is about as toxic as aspirin." Hundreds of Vietnam war veterans in the United States would disagree. Between 1962 and 1971, the U.S. military sprayed over Vietnam about 42 million liters (11 million gallons) of the jungle defoliant Agent Orange, which is composed of a 1:1 mixture of 2,4,5-T and 2,4-D. As early as 1969, newspapers in Saigon reported high rates of miscarriages and birth defects in children born to Vietnamese women living in and near areas that had been sprayed with Agent Orange.

It was not until about 1977 that reports of Vietnam-related illnesses in the U.S. began to show a disturbing pattern of similarity: headaches, nervous disorders, chloracne, tumors, liver disease, and birth defects in veterans' children. No cause-and-effect relationship between 2,4,5-T (and specifically, its dioxin impurities) and health effects has yet been firmly established. An epidemiological (population) study by the U.S. Veterans' Administration (VA) is in the works. Lawsuits by veterans against both the VA and Dow Chemical (and other Agent Orange manufacturers) are pending.

Aspirin and dioxin are hardly comparable. Says American researcher Theodore Sterling: "What appears clear from studies of TCDD's lethality [ability to cause death] is that because the effects of long-term exposures to low levels of TCDD remain undetermined, an acceptable level of exposure for neither man nor test animals can be postulated. If TCDD causes delayed lethality or if continued impingement of TCDD on human organs otherwise causes cumulative effects, or if TCDD accumulates in human tissue, a level of exposure which would be safe for the general population of either man or animal may not exist. Even residues below the current level of detection may be unsafe."

For both 2,4,5-T and PCP, there is considerable discussion as to whether the health effects are due to the pesticides themselves or to the dioxin impurity. When Dow compares 2,4,5-T to aspirin, it can be referring only to the pure product, which rarely exists anyway, since dioxin is an unavoidable contaminant in commercial 2,4,5-T production. However, in comparing differences in effects of analytical grade PCP (with no dioxin) and technical grade PCP (containing the dioxin contaminants found in commercial preparations), research with heifers at the U.S. National Institute of Environmental Health Sciences (NIEHS) and North Carolina State University has indicated that dioxin is to blame.

Says NIEHS pathologist Eugene McConnell: "The results of our study show clearly that the toxicity of PCP in cattle is primarily attributable to its contamination with toxic impurities, either the dibenzodioxins or the dibenzofurans." Still, PCP itself *is* toxic to some degree, affecting the immunity to disease in laboratory and farm animals. (Some researchers attribute even these effects on immunity to dioxin contaminants.)

The primary sources of 2,4,5-T-related dioxin in food would be a result of its use on rice crops and cattle rangeland. PCP-related dioxin in food can arise from the use of "penta" as a wood preservative in barns, feed bins, fences, and wood shavings used as bedding (see *Table 36*). Numerous cases of illness in cattle have been linked to the animals' chewing or rubbing against PCP-treated wood. For example, in 1977

Table 36 / *PCP in Poultry Litter in Ontario*

PCP Residue (range in parts per million)	*Number of Samples*
Nil	5
0.01–0.99	49
1.00–4.99	45
5.00–29.9	27
30.0–139	17

Source: Veterinary Services Branch, Ontario Ministry of Agriculture and Food. Unpublished data, 1980.

---

in Michigan, PCP was detected at levels as high as 1,000 parts per million in the blood from eight cattle herds. The PCP itself contained HCDD, hepta-CDD, and OCDD at 1 to 1,000 ppm. Notes the medical journal *The Lancet:* "Dioxins are indisputably present in cow and calf livers, the latter probably contaminated through milk, since dioxins are readily secreted into milk. In fact, contaminated milk from affected herds now poses serious disposal problems."

In June 1978, the Ontario government was concerned enough about poultry contamination by PCP-treated wood shavings that it warned poultry farmers to stop using treated bedding and to have bedding materials tested for PCP through a free service at the University of Guelph if they were unsure of the quality of the wood. But only about one-third of the farmers have taken advantage of the service and the use of PCP-treated bedding is still widespread because the wood product is cheap and practical. Some chicken and turkey producers have switched to straw, but not enough straw is available generally to allow a wholesale switch. Besides, toxic moulds may infect straw bedding.

2,4,5-TP (Fenoprop) is sometimes used in Canada as a "stop-drop" for apples. Agriculture Canada's research station in Summerland, B.C. has uncovered low (0.036 to 0.097 ppm) residues of Fenoprop in apples, but could not test for dioxin. Most orchardists use alternative stop-drops such as Alar and NAA (naphthalene acetic acid). Still, a regulation under Canada's *Pest Control Products Act* allows TCDD at a level of 100 parts per billion (0.1 ppm) in both 2,4,5-TP and 2,4,5-T.

Few laboratories can test for dioxin because of the danger and expense involved. At one point, the Ontario government's pesticide residue laboratory in Guelph had *one gram* of TCDD on its premises—and that was a big deal. Health and Welfare Canada research laboratories in Ottawa and Ontario Ministry of the Environment laboratories in Toronto are currently being set up to test for dioxin in food at the parts per trillion level.

Because of technical and financial constraints, data on dioxin in Canadian food are very sparse. According to Health and Welfare's Harry McLeod: "We have analyzed a few samples of cow's milk from Alberta. No residues were found with a method capable of screening down to approximately 10 ppt [parts per trillion]. We have not analyzed imported rice as yet for any of the dioxins. A couple of years ago, we analyzed a few samples of potatoes from several provinces with negative results, using at that time a method capable for screening down to approximately 50 ppt."

In 1978, dioxin was detected in two Lake Ontario fish, by a laboratory at the University of Nebraska which has been set up with sophisticated analytical instrumentation specifically for dioxin analysis. The dioxin levels were 6.5 and 4.6 parts per trillion. Possible sources of the dioxin were either the infamous Love Canal waste disposal site (where Hooker Chemicals and Plastics Corporation stored about 60 kilograms [130 pounds] of dioxin) or the nearby Hyde Park site (thought to contain 135 to 900 kilograms [300 to 2,000 pounds] of dioxin), both in upstate New York and both of which contaminate the Niagara River and, therefore, Lake Ontario.

As more and more laboratories become equipped to test for dioxin, the trickle of data on dioxin in food is turning into a steady stream:

● In 1978, Health and Welfare Canada found traces of the hexa and octa forms of dioxin (HCDD and OCDD), both considerably less toxic than TCDD, in chicken livers from the Guelph, Ontario area. (The test results were not made public until December 1980.) Since dioxin is fat soluble, health department officials plan to test chicken fat as well. Alex Morrison, head of the federal Health Protection Branch, insisted that the chicken livers posed no significant hazard to human health. Yet the geographical extent of contamination is unknown and the long-term health impacts of low residues of dioxins in food are poorly understood. The dioxin source in the chicken liver case was PCP-treated wood shavings used as poultry bedding.

● In 1980, extensive tests by the New York State Department of Health and the Ontario Ministry of the Environment uncovered TCDD in several species of fish from Lake Ontario, at levels ranging from 3 to 40 parts per trillion. Douglas Hallett, author of an Environment Canada study that found TCDD in herring gull eggs from Lake Ontario, said, upon release of the fish study, that he personally would not eat fish from the lake.

Data on dioxins in American food are also meagre:

● Studies by Dow Chemical found TCDD in concentrations ranging from 10 to 230 ppt in the tissues of fish downstream from Dow's plant on the Tittabawassee River in Michigan. As a result of these findings, the Michigan Department of Public Health issued a health advisory against eating fish taken from that river and other nearby waters.

● The Environmental Protection Agency has uncovered 40 ppt of TCDD in fish caught in Michigan lakes.

● Samples of beef fat in the meat of cattle grazing on western rangelands that had been sprayed with 2,4,5-T have been found to be contaminated with TCDD, as have samples of human breast milk from Texas and Oregon. (Like other halogenated hydrocarbons, dioxin bio-concentrates in fatty tissues.)

● Preliminary results of an EPA study in 1980 indicated that 13 percent of the American beef fat tested contains dioxin in the ppt range.

● Michigan State University researchers Mary and Matthew Zabik have found that beef liver may concentrate the higher dioxins (HCDD, hepta-CDD, and OCDD, found in PCP) at appreciable levels (although cooking greatly reduced the dioxin residues). They concluded that "the liver clearly would be a public health hazard if dairy cattle are allowed access to pentachlorophenol treated wood."

In 1980, a regulation was passed under the *Food and Drugs Act* stipulating that a food is adulterated if it contains chlorinated dibenzo-p-dioxins. In other words, *any* amount of *any* of the dioxins is now illegal in Canadian food. However, this no-residue standard is under review as more becomes known about the health effects (if any) of minute quantities of dioxins.

PCP itself (not the dioxin contaminants) is currently permitted at a level of 0.1 part per million in all foods. A 1980 Agriculture Canada study found 5 of 170 poultry samples taken in spot checks exceeded the 0.1 ppm tolerance. (The meat was not checked for dioxins.) Agriculture department tests of beef and pork have not uncovered any PCP residue.

Dioxin levels in pesticides are much lower now than they were in the early days of 2,4,5-T and PCP use. When manufacturers first began synthesizing 2,4,5-T after World War II, its TCDD content was sometimes hundreds of times higher than it is today. The TCDD in Dow Chemical's 2,4,5-T was reduced from 1.0 ppm in 1970 to 0.03 ppm (30 ppb) in 1976, well below the permitted 100 ppb. However, it costs more to produce pesticides with lower dioxin. Recently, Dow introduced a new wood preservative, Dowicide EC7, in which dioxin levels had been reduced two orders of magnitude by using more expensive production techniques. But the new product was found to be non-competitive against rival, non-decontaminated brands, and Dowicide EC7 has now virtually disappeared from the marketplace.

It is clear that alternatives to dioxin-containing pesticides and preservatives must be used wherever possible. The herbicide 2,4,5-T has been banned or restricted in some Canadian provinces. The replacement of PCP by compounds containing arsenic, cadmium, or creosote is not much of an improvement. One hopeful sign is a means of preserving wood developed by the U.S. Department of Agriculture, using a water repellent wax in mineral spirits plus 10 percent resin, and no PCP.

In addition, stepped-up precautions must be taken to ensure that dioxin-containing industrial wastes are properly disposed of. Dioxins may be present in the waste products from chlorination of benzenes, phenols, and polyphenyls. In addition, dioxins are formed as a by-product of the incineration of municipal solid waste, such as waste PCP-treated wood.

Other Organics

A number of other organic chemicals—some halogenated hydrocarbons, some not—can find their way into our food. In the examples given below, the contamination is not widespread, but where it does exist, it can be serious.

● Mirex

Like DDT, PCBs, and dioxin, mirex is a halogenated hydrocarbon. Also akin to these other chlorinated organic chemicals, mirex is persistent in the environment, accumulating in fatty tissues of living organisms. Mirex is an insecticide for the control of fire ants in the southern United States and a fire retardant in plastics and other synthetic products.

Mirex is no longer used in Canada at all. (A substitute, Dechlorane Plus, is used, but the safety of this compound is also suspect, since its chemical structure is related to that of the insecticides aldrin/dieldrin, chlordane, and heptachlor.) The only significant input of mirex to Canadian food has been its detection in Lake Ontario fish, due to improper containment facilities at the Hooker Chemicals and Plastics Corporation in Niagara Falls, New York, and the Armstrong Cork Company in Volney, New York. Neither company handles mirex any longer.

Mirex has been shown to be toxic to non-target species such as algae and shellfish; to affect fish behaviour; to reduce hatchability of eggs and inhibit survival of the young of some birds; and to induce tumours in some laboratory animals.

The Ontario Ministry of the Environment and Ministry of Natural Resources surveyed about 1000 Lake Ontario fish in 1976. Mirex was found in all of the 18 species of fish tested. Most species were under the 0.1 part per million guideline for human consumption set by the Ontario and U.S. governments for mirex in fish, but this maximum permitted level was exceeded in some brown trout, coho salmon, smelt, white perch, and yellow perch. As in the case of PCBs and DDT, warnings have gone out to anglers to restrict consumption of certain species and sizes of Lake Ontario fish because of mirex contamination.

Very low levels of mirex have been found elsewhere in Canada. Some wild birds in the Maritimes, Prairies, and central Canada contain mirex, probably as a result of bird migration to the southern U.S. (remember the fire ants?) rather than local contamination.

● Benzene

In Canada, benzene ranks second only to ethylene in the production of industrial organic chemicals. It is not a halogenated hydrocarbon. It is used as a solvent and as a starting material for the manufacture of a wide variety of products, such as plastics and pesticides. It is also present in gasoline. It is a very stable, persistent substance.

The greater and more serious benzene exposure is occupational rather than environmental. Health effects in work places containing benzene vapours include nervous system disorders and leukemia.

According to a study by Health and Welfare Canada, the total intake of benzene by urban dwellers from non-occupational sources is about 125 milligrams per year, of which 90 milligrams come from food. However, data on benzene in food are extremely limited. Benzene is thought to occur naturally in fruits, vegetables, nuts, dairy products, and eggs. The U.S. National Cancer Institute found benzene at 2.1 parts per million in eggs, 120 parts per billion in Jamaican rum, and 2 to 19 parts per billion in beef.

Two possible sources of benzene in food are accumulation in drinking water (benzene has been found in the water supply of Niagara-on-the-Lake, Ontario) and fallout of benzene from vehicle exhaust onto crop lands. According to Consumer and Corporate Affairs Canada, the only consumer products which have been found to contain benzene are felt tip markers, tire repair kits, shellac, and ballpoint pens. Other products containing benzene include solvents, grease cutters, paint removers, wax removers, lacquers, and pesticides.

● PBBs

Aside from the 1973 accident in Michigan (mentioned at the beginning of this chapter), the repercussions of which are still being felt, PBBs are not a common food contaminant. PBBs have not been used in Canada since 1975, and their use was banned in this country in 1978. Shipments into Canada of Michigan-produced meat and dairy products were halted temporarily after the accident, and Canadian federal authorities continue to monitor Michigan products for PBBs. No PBB residue is permitted.

We'll conclude with the words of Yale University biologist Arthur Galston: "Undoubtedly, the employment of suitably sensitive tests for the detection of toxic impurities, as well as the use of proper screens for mutagenicity, teratogenicity, and ecological damage will uncover yet other suspect compounds.... The unfortunate fact is that many of the currently employed compounds were introduced into wide use without proper toxicological evaluation. It will be agonizing and expensive for industry, professionals, and practitioners to have to withdraw or modify some of their long-favored compounds, but the public welfare may demand such a step."

What You Can Do

● Severely restrict fishing trips in waterways known to be polluted with industrial chemicals and pesticides. New mothers, other women of childbearing age, and children should eat such catches only occasionally.

● New mothers in heavily industrialized areas and those occupationally exposed to toxic chemicals should have their breast milk tested for PCBs and other pollutants. Insist that your health department perform this test (without charge).

● Don't use the wood preservative pentachlorophenol (PCP) or the herbicide 2,4-D, which may contain dioxins. (2,4,5-T, which contains the highly toxic dioxin TCDD, has no home-and-garden uses.)

10 Metals as Food Pollutants

- In 1975, the year that the Reed Paper Company stopped dumping mercury into the Wabigoon River at Dryden, Ontario, the level of methyl mercury in the blood of 11 native fishing guides working in the area averaged 136 parts per billion. All 1,200 Indian people on the two reserves downstream from Reed, Grassy Narrows and White Dog, have been affected by the mercury to some degree, due to consumption of contaminated fish. Mercury had been used in the production of chlorine, needed for paper production.

- In September 1979, the Ontario Ministry of the Environment ordered the city of Kingston to stop hauling sludge from its sewage treatment plant to nearby farmland for purposes of fertilization, because the sludge contained potentially dangerous levels of chromium. The chromium was believed to have originated at the Alcan (Canada) Products Ltd. plant in the city.

- In October of the same year, the Ontario division of the Loblaws Ltd. supermarket chain pulled its "no-name" canned pork luncheon meat off store shelves after the Manitoba health department warned that province's citizens not to eat the meat. The reason? Several cases of metal contamination by lead and tin from the cans had been found.

This chapter will show how metals contaminate the food supply and the ways in which these substances are toxic. There will be separate sections on three metals that are particularly important food contaminants—lead, mercury, and cadmium—with information on levels of these metals in the Canadian food supply. The intake of metals in the overall diet will be discussed, as well as the ways in which their intrusion into food can be minimized through government regulation and changes in industrial processes.

Metals, Metals Everywhere

Food pollution by metals is almost as old as time, but time certainly hasn't solved the problem. Ancient contributors of metals to foods, such as the lead pipes of the Roman empire, have been replaced by modern sources, such as cadmium in livestock drugs and arsenic in pesticides. Indeed, as the U.S. National Academy of Sciences put it in 1971, "The two groups of chemicals that appear to offer the greatest danger through promiscuous release to the environment are the heavy metals and halogenated hydrocarbons."

Heavy metals are also sometimes referred to as trace metals (as opposed to the precious metals such as gold and silver) or trace elements, because under normal, uncontaminated conditions, they occur in very low concentrations ("traces", from parts per billion to virtually undetectable amounts) in nature (see definition in the Glossary, Appendix B). Unlike synthetic organic compounds—pesticides, PCBs, and so on—heavy metals are not made by humans. What technology has done is redistribute and concentrate metals to the point that they pose pollution problems.

Heavy metals can appear in food either because of the natural levels found around us or as a result of inadequate control on the use of metal-containing substances in agriculture, industry, fuel combustion, and other activities. It is difficult to determine what are "natural" and what are "contaminated" levels in modern foods, but in most cases of gross contamination, a man-made culprit can be identified, as the examples opening this chapter indicated.

Since heavy metals are often present in minute quantities and exert their effects at these trace levels, the sensitivity of laboratory equipment and procedures used to measure metals in food is of critical importance. Leon Russell Jr. of Texas A & M University has said, "As the sensitivity of analytical procedures has increased, so has our understanding of metals and their action on biological systems."

The characteristic of heavy metals that sets them apart from organic pollutants is that some of them are, at very low levels, actually beneficial, in fact essential, to living organisms. The 10 trace elements that have been established as essential to proper nutrition in mammals are iron, copper, manganese, zinc, cobalt, iodine, molybdenum, selenium, chromium, and tin. These have been called the nutritional elements or "micro-nutrients," in contrast to metals such as lead, mercury, and cadmium, the "toxic elements" that are harmful in any concentration.

But even the micro-nutrients can be toxic if present in amounts greater than those required by the organism. A good example is selenium; it is required in the diet of livestock at a minimum level of 0.04 parts per million and is beneficial up to 0.1 ppm, while at levels above 4 ppm, it becomes toxic to animals. Although some forage plants are high in selenium, farmers have suffered greater losses because of selenium deficiencies than selenium toxicities. For many other micro-nutrients, the difference between essential and toxic concentrations is much greater.

However, it would be misleading to define safe or unsafe levels of metals in foods without considering interactions among metals. Notes Eric Underwood of the University of Western Australia: "The absorption, utilization and excretion of many trace elements, and therefore their physiological, pharmacological, or toxicological actions within the cells and tissues of the animal body, are greatly influenced by the extent to which other elements or compounds with which they interact are present or absent from the diet and from the body itself. Thus there is no single minimum need of an essential trace element and no single safe dietary level of a toxic element. There is a series of such minimum needs or maximum tolerances depending upon the chemical form of the element [elemental form, inorganic salt, organic compound], the duration and continuity of intake, [also, the mode of entry into the body—inhalation, absorption through the skin, or dietary intake] and the nature of the rest of the diet, especially the amounts and proportions of other interacting elements and compounds."

For example, the micro-nutrients zinc and selenium are known to interact in the body with cadmium and mercury, respectively, so that the adequacy of the intake of zinc or selenium can be influenced by the degree of exposure to cadmium or mercury. Similarly, the toxicity of the cadmium and mercury can be influenced by levels of the zinc and selenium. Therefore, concludes Underwood, "It cannot be stressed too strongly that metabolic interactions among the trace elements can be so powerful that environmental studies involving single elements can lead to dangerously erroneous conclusions."

Metals are toxic in many and varied ways. Some attack nerve tissue by blocking the action of certain key enzymes. More often, metals may exert their action indirectly, through destruction of tissues and processes in important organs such as the liver and kidneys. Some metals are at least suspected of being carcinogenic and/or teratogenic. Toxicity of heavy metals in animals is affected by the level of ingestion, age, species, sex, physical condition, nutritional adequacy of the diet, and the capacity of different tissues to store the metals.

Metals move in curious ways through the environment:

● Because they are persistent chemicals, they are well known for their potential to accumulate in living things.

● Some plant species can take up relatively massive amounts of certain metals and show no toxic effects, while others are extremely sensitive. For example, lettuce and celery are particularly tolerant of metals in their tissues.

● Moving up the food chain, we find that metals migrate to certain organs in animals. Metals concentrate in liver and kidney, but not due to the fat content (the material that attracts organochlorine compounds) of these organs. Rather, the functions of these organs in breaking down chemicals unneeded by the body and preparing them for excretion in the urine make the liver and kidney good metal accumulators.

● Fish are efficient accumulators of heavy metals from the aquatic environment. For the purpose of scientific research, they are therefore useful "indicator species." Studies of fish tell us a great deal about long-term changes in the metal levels around us.

● The absorption of a metal (consumed in the diet) from the digestive tract into body tissues is aided by the acidity in the human gut but is restricted by the low permeability (tendency to let certain chemicals pass through) of the walls of the gut to many metals.

Sources and Toxicities of Metal Pollutants

The sources (though not all industrial uses) of toxic heavy metals found in food and the kinds of toxicity these chemicals produce are outlined in *Table 37*. Generally, metals in food have three sources: environmental pollution of air and water leading to pollution of food; accidental inclusion during food processing; and contamination from food containers during storage.

Symptoms of metal poisoning can be so generalized and vague that an incorrect diagnosis is possible. For example, mercury poisoning mimics some of the characteristics of drunkenness, and several metals cause gastroenteritis (see *Table 37*), an ailment which can have numerous causes. This is a common problem in treating environmentally caused illness. According to Alan MacGregor of the Massachusetts Institute of Technology, "The possibility exists that many unusual deaths which periodically occur in our society may have been caused

---

Table 37 / *Heavy Metal Contaminants in Foods: Sources and Toxicities**

Metal and Symbol	Sources	Toxicities
Lead (Pb)	motor vehicle exhaust (from use of tetraethyl lead in gasoline); solder used to seal the seams on metal food cans; agricultural pesticides (e.g., lead arsenite as a defoliant in potatoes); fertilizers (especially those compounded from sewage sludge); fallout from lead smelters and lead battery operations; lead-glazed food storage, cooking and eating utensils; lead-based inks used on food packages; paints; water pipes; fallout from coal-fired power plants; lead caps on wine bottles; lead shot (for hunting); waste oils (automotive, industrial) spread on dirt roads to suppress dust	gastroenteritis; anemia; lead paralysis and colic; aggravation of hyperactivity; behavioural and psychological effects (at low doses); children are especially sensitive to lead (e.g., brain damage, possibly mental retardation)

* Toxicities refer to effects in laboratory animals and, where known, to effects in man.

Metal and Symbol	Sources	Toxicities
Mercury (Hg)	water pollution from the chlor-alkali industry (uses Hg to produce chlorine and caustic soda) and the pulp and paper industry (used to use Hg as a slimicide); agricultural fungicides (methyl mercurial seed dressings for grain—now banned in Canada; phenyl mercuric acetate to prevent scab in apples); horse manure (found in compost in mushroom culture); antifouling paints (for boats and other aquatic uses); electrical and plastic industries; fallout from coal-fired power plants; fertilizers (especially those compounded from sewage sludge); broken thermometers; mining, milling and smelting of ores	severe gastroenteritis and kidney inflammation, even death (from high acute doses); irreversible brain damage (from chronic exposure); nervous system effects such as tremors (especially in mercury miners and those employed in the felt hat industry); tunnel vision, hearing difficulties, numbness, speech impairment, loss of coordination, and paralysis; suspected teratogenicity; most toxic and most common effects are from organic forms (e.g., methyl mercury), not inorganic forms (e.g., mercury salts)
Cadmium (Cd)	pollution from electroplating industries; agricultural use of sewage sludge as fertilizer; superphosphate fertilizers; cadmium-based fungicides (mostly for turf); livestock drugs; cadmium-plated food storage, cooking and eating utensils (no longer in production); metal alloys; cadmium as a contaminant in any use of zinc (they occur together in ore bodies), such as galvanized utensils; wearing of motor vehicle accessories (tires, bumpers); cadmium-based glazes on ceramic cookware and eating utensils; cadmium-based inks used on food packages and cadmium stearate used as a plastic stabilizer; discarded articles littered on farmland (painted surfaces, batteries, etc.); motor oils; fossil fuel combustion; solders and pipes	severe gastroenteritis (at high acute doses); "itai-itai" disease (includes kidney damage and severe bone disorders); chronic lesions (testicular tissue death, birth defects, cardiovascular hypertension); lung and kidney disease; suspected teratogenicity and carcinogenicity; most acutely toxic heavy metal
Arsenic (As)	agricultural pesticides (e.g., sodium arsenite as a potato	gastroenteritis, cardiovascular collapse, cirrhosis of the liver,

Metal and Symbol	Sources	Toxicities
	defoliant; lead arsenate and calcium arsenate in fruit orchards) most of which have been replaced by organic pesticides; livestock drugs (arsenicals such as arsanilic acid and 3-nitro-4-hydroxy phenyl arsonic acid used as growth promotants in poultry and swine); fallout from coal-fired power plants; gold-mining wastes; fallout from lead smelters; fertilizers	kidney disease and jaundice (at high acute doses); loss of weight and hair and skin lesions, skin cancer in vine-yard workers, liver cancer, paralysis of legs, heart disease (from chronic intoxication)
Tin (Sn)	tin cans used to package foods; use of large amounts of nitrate fertilizer (since high-nitrate foods promote "de-tinning" of tin cans used to package some foods)	regarded as a low-toxicity metal; gastroenteritis at high acute doses
Nickel (Ni)	organic fertilizers (especially those compounded from sewage sludge); fallout from nickel smelters; wearing of nickel-containing vehicle engine parts; discharges from electroplating plants; stainless steel food processing, storage and cooking surfaces; nickel-based catalysts used in the hydrogenation of edible oils; fallout from coal-fired power plants; diesel and fuel oil	a low-toxicity metal
Copper (Cu)	agricultural fungicides; fertilizers (especially those compounded from sewage sludge); copper food storage and cooking utensils; fallout from copper smelters	gastrointestinal disorders at high doses
Zinc (Zn)	fertilizers (especially those compounded from sewage sludge); galvanized food storage and cooking utensils	gastroenteritis (from high acute doses); not considered a toxic hazard to man
Chromium (Cr)	stainless steel food processing, food storage and cooking surfaces; chrome plating on motor vehicles; chromium-based paints and inks; chro-	considered a low-toxicity metal at levels found in food, if in the Cr^{+3} form; highly toxic in the Cr^{+6} form; main hazard is in occupational settings (der-

Metal and Symbol	Sources	Toxicities
	mium-based catalysts used in the hydrogenation of edible oils; fertilizers (especially those compounded from sewage sludge)	matitis, injury to respiratory tract including lung cancer, and effects on the nervous and digestive systems)
Selenium (Se)	natural occurrence primarily; used in glass manufacturing; found in selenium-based colours for paint, plastics and ceramics; rubber additive	inhibits some enzymes; suspected carcinogenicity; in livestock: blind staggers disease, decreased growth, hair loss, abnormal hoof development and death from ingestion of selenium-accumulating plants

Sources: See source list for this chapter (Appendix C).

---

by toxic chemical poisoning, but were never diagnosed as such because the doctor or coroner failed to recognize symptoms of such poisoning."

Historically, the worst effects on human health from heavy metals have been due to occupational rather than consumer (general environmental) exposure. As much emphasis, if not more, should be placed on curbing metal release in the workplace as on the dinner plate.

The heavy metals of greatest concern with respect to food contamination are lead, mercury, and cadmium. Each is discussed later on, after a look at sewage sludge and acid rain as heavy metal sources.

Contamination by Circuitous Routes

Two sources of heavy metals in food are particularly interesting because of the indirect channels by which they come to contaminate what we eat. One—the use of sewage sludge as fertilizer on agricultural land—is a proven source of metals for food crops. The other—acid rain—is at present only hypothetically implicated in the process of metal contamination of food.

● Sewage Sludge (Recycling Gone Sour)

The process of treating sewage produces unsavory goop called sludge. Most of the metal content of the raw (untreated) sewage ends up, after treatment, in the sludge, as do more desirable substances such as phosphorus and nitrogen, particularly the latter. The problem of disposing of sludge has been "solved" by using it to fertilize farmland.

Notes Peter Hannam, past president of the Ontario Federation of Agriculture: "Fertilizer production, particularly nitrogen, is a large user of energy. Nitrogen is an essential nutrient for all plants except legumes, which produce their own. Farmers are finding several ways to reduce reliance on nitrogen fertilizers. They are making better use of livestock and poultry manure—good sources of nitrogen. They are beginning again to include more legumes (clover, etc.) in crop rotations. There is a tremendous potential source of nitrogen in urban sewage sludges. Some sludges are being spread on farmland now, although many sewage sludges contain too much heavy metal. These could also be utilized if those metals could be eliminated at source."

Herein lies the rub—a great recycling idea gone a touch sour. While the water effluent of sewage treatment plants is relatively clean, the sludge, especially in cities with heavy industries, can be messy. Metals which most often cause problems in sludge are cadmium, zinc, and copper (and, to a lesser extent, chromium, lead and nickel), occasionally reaching levels that can lead to unacceptably high metal concentrations in crops grown on land to which sludge has been applied. The higher the acidity of the soil (see pH in Glossary, Appendix B) and the lower the humus (organic matter) content, the easier it is for crops to take up metals from the soil.

According to a 1979 report by the National Research Council of Canada, cadmium, more than any other metal, may be responsible for eliminating various sludges from possible use on farmland. The cadmium content of some Ontario sewage sludges is shown in *Table 38* and uptake of cadmium by corn grown in sludge-treated soil is given in *Table 39*. Maximum desirable levels for all metals in sewage sludge to be spread on farmland are given in *Table 40*.

Table 38 / *Cadmium Content of Some Ontario Sewage Sludges, 1973 to 1974*

Source of Sludge	Cadmium Concentration (parts per million)
Newmarket	4.0–5.3
Ottawa	11.6
Toronto	16.5–37
Sarnia	99

Source: National Research Council of Canada (Associate Committee on Scientific Criteria for Environmental Quality). *Effects of Cadmium in the Canadian Environment.* NRCC Publication No. 16743. Ottawa: National Research Council of Canada, 1979.

Table 39 / *Uptake of Cadmium by Corn from Guelph, Ontario, Soil Treated with Sewage Sludge*

Sludge Application Rate (kilograms per hectare)	Cadmium Levels in Corn (parts per million)	
	Leaves	Kernel
0	3.3	0.3
27	3.0	0.6
54	5.3	0.8
108	11.6	1.0

Source: National Research Council of Canada (Associate Committee on Scientific Criteria for Environmental Quality). *Effects of Cadmium in the Canadian Environment.* NRCC Publication No. 16743. Ottawa: National Research Council of Canada, 1979.

Table 40 / *Metal Contents of Sludge Appropriate for Farmland Application*

Metal	Maximum Appropriate Content in Sludge (parts per million)
Zinc	2,000
Copper	800
Nickel	100
Cadmium	10*
Boron	100
Lead	1,000
Mercury	15

* 0.5% of zinc content

Source: R.L. CHANEY, "Crop and Food Chain Effects of Toxic Elements in Sludges and Effluents." *In* Proceedings of the Joint Conference on Recycling Municipal Sludges and Effluents on Land. Washington, D.C.: The National Association of State Universities and Land-Grant Colleges, 1974.

Ontario is the Canadian province with both the heaviest concentration of industrial activity and the most comprehensive program for the application of low-metal sewage sludge onto farmland. Of all the sludge produced in the province, 34 percent is applied to land and 40 percent is incinerated. The rest ends up in "sanitary" landfill sites.

Under the direction of Gordon Van Fleet of the Ontario Ministry of the Environment, a three-year phase-in program for use of safe sludges on farmland is being implemented. Criteria for deciding if a sludge is safe include its content of 11 metals, its acidity (the pH must be equal to or greater than 6), and its ratio of nitrogen to specific metals (the ratio must be at a certain minimum for each metal).

Van Fleet expects about 70 percent of the sludges now going on farmland to be low enough in metals to continue to be disposed of in this way. There are no practical methods for removal of heavy metals from sludges; the solution is to reduce metal levels "at source"—that is, in the industry or other activity that is polluting the water to begin with.

Fertilizing farmland with sewage sludge makes good sense—as long as we know what we're dealing with. As Rufus Chaney of the U.S. Department of Agriculture has put it, "Until we have more field knowledge of toxic element injury and accumulation from sludges and effluents, we will not be sure we are dealing with all of the potential problems."

• Acid earth from acid rain

Rain and snow acidify due to their combination with the air pollutants sulphur dioxide and nitrogen oxides. Acid rain's ability to lower the pH of lakes in Sweden, Ontario and the Adirondacks in the northeast U.S. is already well known. What is less understood is the likelihood that, if gone unchecked, acid rain could lower the pH of agricultural soils to such an extent that heavy metals in these soils become more available for uptake by crop plants.

If acid rain does produce acid soil, then liming to make the soil more alkaline is one antidote. (Clearly, this would not be a solution for the millions of acres of forests that might also be affected by acid rain.)

"To say that acid precipitation is not an immediate problem on agricultural lands is probably true if you are just looking at the soil and making the assumption that all farmers will follow good management practices," said Peter Rennie of the Canadian Forestry Service in a 1980 report on acid rain's effects on soil in *Harrowsmith* magazine.

"The calcium reserves of soils may be large, and normal soil amendment overcomes the small losses caused by acid rain. But in many farming areas, management practices are not very perfect and, in fact, adequate liming is *not* being done. Acid rain may also do serious damage to the foliage of agricultural crops [not related to heavy metal problems], something that no amount of liming can prevent."

Lead

It has been suggested that the fall of the Roman empire was attributable in part to lead poisoning. Acidic foods such as wine were often stored in earthenware containers that had a lead-based glaze. Relaxation of prohibitions against the drinking of wine by women of the aristocracy (in addition to water pipes containing lead) may have resulted in sterility, which substantially reduced the size of the next aristocratic generation.

In *Toxicology of Pesticides,* physician Wayland Hayes, Jr., states that organochlorine pesticides are relatively safe if one compares their residues in food with those of lead. Daily ingestion of three milligrams of lead—only about nine times as much as ordinarily occurs in the diet (rather than any order of magnitude such as 100 or 1000 times)—produces symptoms of lead poisoning.

Lead is everywhere. Since the Industrial Revolution, and particularly since the mass introduction of automobiles using gasoline containing tetraethyl lead as an antiknock additive, levels of lead in the environment have increased markedly.

The two sources of lead that now figure most prominently in food contamination are leaded gasoline and lead-soldered tin cans for processed foods:

● It is estimated that 40 percent of all farmland in the United States receives lead fallout from the millions of motor vehicles that ply the nation's highways and biways. The move towards unleaded gasoline will prevent further contamination, but because lead is chemically stable, the sins of the past will be visited upon the future.

● Fully half of the lead in our diet could be eliminated if we stopped packaging food in cans in which the seam is sealed with lead solder.

Like mercury, lead is both persistent and capable of bio-concentrating in food chains. Fortunately, though, it is not easily taken up by plant roots. In 1979, I took to a University of Toronto laboratory samples of produce from my vegetable garden, located only about 100 metres from Highway 401, the mega-road which crosses the north end of Toronto. Tests showed that while lead levels in the soil were unnaturally high—of the order of 120 parts per million (uncontaminated levels in soil are 1 to 50 ppm)—the crops themselves contained levels of lead close to what one would expect to find in crops grown in uncontaminated soils—1 to 8 ppm (uncontaminated crops: 0.05 to 3 ppm).

However, at extremely high soil lead levels, crops can become dangerously tainted. People living near the Canada Metal Co. Ltd. lead smelter in Toronto have been warned not to grow vegetables in their gardens.

The higher the acidity of the soil (i.e., the lower the pH), the more available lead and other heavy metals become to plant roots. Also, once in the plant, lead remains in the roots, so root crops would be of primary concern. Some crops have a greater tendency to pick up lead than others; radishes, spinach, carrots, and lettuce are particularly high accumulators.

Only about 10 percent of the lead we consume in our food is actually absorbed into our bodies; the majority is excreted in urine and feces. However, the situation is much different in children. Not only do their bodies absorb a higher proportion of the lead consumed—as high as 60 percent—but they are intrinsically more susceptible to lead poisoning because of their rapid metabolic rate and small body size. *Any* person whose diet is low in calcium, iron or protein, and/or abnormally high in vitamin D will suffer greater effects from lead in food.

While the health effects of acute overexposure to lead are well known (see *Table 37*), what is now becoming evident is that lead can have psychological effects at very low doses. In a March 1979 paper in the *New England Journal of Medicine,* Herbert Needleman and his colleagues at the Mental Retardation Research Center of the Children's Hospital Medical Center and Harvard Medical School reported: "Lead exposure, at doses below those producing symptoms severe enough to be diagnosed clinically, appears to be associated with neuropsychologic deficits that may interfere with classroom performance."

Tests involving 270 Massachusetts school children found that those with the highest lead levels in their teeth performed much more poorly in intelligence tests, especially the verbal items, in three measures of auditory and verbal processing, in attention span, and in most behavioural characteristics measured by teachers (easily distracted, disorganized, hyperactive, easily frustrated, etc.).

The authors concluded that "the impaired function of children with high lead levels, demonstrated in the neuropsychologic laboratory, mirrored by disordered classroom behavior, appears to be an early adverse effect of exposure to lead. Permissible exposure levels of lead for children deserve reexamination in the light of these data."

The lead intake from food of the average Canadian is about 140 to 160 micrograms per day, although the level varies significantly with food preferences and geographical location. Levels in Canada are about the same as those in the United States, but only half of those found in food in Japan. Generally, it is canned goods—evaporated

---

Table 41 / *Lead in the Diet, Canada and the United States, 1971 to 1972*

Food Group	Average Total Lead Intake (micrograms per person per day)	
	U.S.A. (1972)	Canada (1971)
Milk and dairy products	21	22
Meat, fish, and poultry	20	18
Cereals	25	22
Potatoes	10	11
Leafy vegetables	7	2
Legumes	20	5
Root vegetables	5	2
Garden fruits	12	8
Other fruits	16	36
Oils and fats	3	2
Sugar	5	10
Drinks	15	1
Total	159	139

Sources: U.S.A.: A.C. Kolbye et al. "Food Exposures to Lead." *Environmental Health Perspectives,* May, 1974, pp. 65–74.

Canada: T.D. LEAH, *Environmental Contaminants Inventory Study No. 3: The Production, Use and Distribution of Lead in Canada.* (Report Series No. 41, Inland Waters Directorate) Ottawa: Environment Canada, 1976.

milk, canned infant formula, canned meats and fish, and canned fruits and fruit juices—which have the highest lead levels. A breakdown is given in *Table 41*. Compare the totals—139 ppm in Canada and 159 ppm in the U.S.—to the Food and Agriculture Organization/World Health Organization tolerable daily intake of 429 micrograms per person for adults. Lead in our food, therefore, is almost half the maximum level that the body can cope with.

Research at the New York State Department of Health by Douglas Mitchell and Kenneth Aldous has examined safe lead levels in food for children. If a child is fed canned products averaging 300 micrograms of lead per litre (50 percent above the mean lead level found in their studies), then it would take only *0.17 litre per day* to exceed the World Health Organization's maximum daily permissible intake of 50 micrograms per day for a 10–kilogram child.

Lead Again: The Can Story

"Since Nicolas Appert made his brilliant discovery, the process which he suggested for the first time in 1814 involving the use of tin cans as air-tight containers which are heated to preserve foodstuffs has continued to expand," writes Claude Boudene of the Laboratory of Toxicology at the University of Paris-South. "Nowadays, tinned foodstuffs sometimes account for a substantial proportion of modern man's diet. It is therefore important also to consider the inevitable metal contaminations produced by the very nature of this type of container."

While the lead present as a component of glass and plastic food packages is tightly bound up in the chemical structure of those materials, poisoning from lead in tin cans is like an accident waiting to happen. While tin is also a toxic metal, it is both considerably less harmful than lead and not readily available to contaminate food, since most tin cans are lacquered on the inside to protect them from corrosion by foods.

Lead contamination of tinned goods arises from the use of lead solder in the seams of the cans, which began in 1930. The solders used before World War II consisted of alloys containing one-third to one-half tin and two-thirds to one-half lead. But the need to conserve tin during the war and the cheapness of lead as well as its malleability, led to the use of new solders with only 2 percent tin and 98 percent lead.

Boudene states: "The use of virtually pure lead in this solder was justified on the grounds that it is located outside the can. However, it frequently contains burrs and, in any case, there exists a meniscus or thread of solder inside the can which is, therefore, in contact with the foodstuffs."

The criteria which promote the migration of metals from the can into the food are as follows:

● the acidity or pH of the food — This is the most important factor. Metals dissolve more easily when the acidity is high (see *Table 42*). So, lead can be expected to be particularly high in canned fruit, fruit juices, and tomatoes. The longer a canned food is stored before opening and eating it, the higher the lead in it will be.

Table 42 / *The Acidity of Some Foods*

pH*	Food⁺
2	lemons, vinegar, grapefruit, apples
3	tomatoes, beer
5	beans
6	corn, milk
7: neutral	
8–14: alkaline	

* a measure of acidity (see Glossary, Appendix B); the lower the pH value, the higher the acidity

⁺ not packaged in cans

Source: Harrowsmith Staff Report. "The Acid Earth." *Harrowsmith,* no. 27 (1980), pp. 32–41, 93.

● the presence of oxygen — A can left open in the refrigerator is exposed to the oxygen in the air, which accelerates the movement of metals out of the can into the food.

● the nitrate content of the food — High-nitrate fertilizers often used in agriculture give rise to the high nitrate levels in foods. Nitrate causes "de-tinning" in canned foods, an effect which increases with time in storage.

There has been plenty of research to confirm that lead levels in canned goods can be high and have been caused by the cans, not some other source of lead. Some Canadian figures are shown in *Table 43*. Note, in particular, the difference between fresh and canned tomatoes.

Table 43 / *Lead Contents of Fresh and Canned Food Products in Canada, 1974 to 1975*

	Product	Lead Content (parts per million)
1974	Milk (fresh)	0.044
	Condensed milk	0.18 (0.097)*
1975	Milk (fresh)	0.040
	Condensed milk	0.10 (0.088)*
	Apples (fresh)	0.16
	Apple juice	0.22
	Tomatoes (fresh)	0.04
	Tomatoes	0.31
	Tomato juice	0.30

* These figures represent the calculated contribution of lead from the milk in the concentrated product.

Source: D.E. COFFIN and W.P. MCKINLEY (Health and Welfare Canada). Unpublished data on chemical contaminants in the Canadian food supply, presented at the Fifth International Congress of Food Science and Technology, Kyoto, Japan, September, 1978.

New York State researchers Douglas Mitchell and Kenneth Aldous compared lead levels in canned and bottled food products. They found that the average lead concentration in the canned foods was 167 micrograms per litre, while the average in the bottled foods was only 42 micrograms per litre.

Of particular concern is the contamination of baby foods by lead from cans, since, as mentioned earlier, lead is more toxic to infants and children than adults. When queried recently about the switch from cans to jars for baby food in the late 1960s, an official at H.J. Heinz Co. in Leamington, Ontario stated that the change came about after market studies indicated that parents preferred to be able to see what they're buying for their babies by purchasing foods in glass jars.

That may be true, but lead also figured prominently in the baby food industry's voluntary switch to glass. Health and Welfare Canada scientists D.E. Coffin and W.P. McKinley of the Health Protection

Branch state: "Particularly high lead levels have been found in some canned baby food products. To avoid high lead exposures to young children, most Canadian and United States baby foods have for the past few years been packed in glass jars.... The only two types of products packed exclusively in cans are evaporated or condensed milks [and ready-to-serve formulas].... and juices and drinks."

Some figures are given in *Table 44*. The high lead contents of the beverages arise from the acidity of fruit juices, which is at pH values of 2.7 to 3.9. This acidity, combined with the high seam/volume ratio of the small cans used to pack these baby foods, would explain the high lead contents.

Table 44 / *Lead Content of Canadian Baby Foods, 1975*

Product	Lead Content (parts per million)
Evaporated milk (canned)	0.06
Infant formulas	0.07
Meat and meat dinners	0.03
Vegetables	0.02
Fruits and desserts	0.04
Juices and drinks (canned)	0.26*
Dry cereals	0.10

* individual values up to 0.85 ppm

Source: D.E. COFFIN and W.P. MCKINLEY (Health and Welfare Canada). Unpublished data on chemical contaminants in the Canadian food supply, presented at the Fifth International Congress of Food Science and Technology, Kyoto, Japan, September, 1978.

A 1980 report in the journal *Science* explained, in minute detail, the way in which inaccuracies in laboratory techniques for measuring lead in food have obscured the difference between lead contributed by cans and "normal" ("background," "natural") levels of lead in food. Dorothy Settle and Clair Patterson of the California Institute of Technology studied lead in tuna (albacore) and have exposed "the erroneous belief, held by the FDA [U.S. Food and Drug Administration], that cans soldered with lead elevate lead levels in foods only a few times above so-called normal levels."

Settle and Patterson found that there was a 10,000-fold difference between the lead concentration that is thought to have been present in prehistoric tuna muscle and that in tuna packed in lead-soldered cans. About 99.5 percent of this contamination originates from lead solder. The rest, considered "natural" lead by many laboratories, results largely from lead pollution of the oceans by atmospheric fallout of lead from gasoline exhaust.

U.S. laboratories quoted in the *Science* paper found lead in canned tuna to the tune of 0.7 to 1.4 parts per million. Health and Welfare Canada reports that lead levels in Canadian canned tuna average 0.3 ppm; the range of values is from 0.02 to 1.16 ppm.

Other than cans, the only types of food packages or utensils that contribute measurable amounts of lead to food are ceramic containers and dinnerware. Lead is used in the glazes applied to china, pottery, stoneware, and enamel ware. The best glazes are the lead silicates, and although lead bisilicate is not very soluble in the dilute acids found in acidic foods, its solubility can be altered drastically by improper formulation or application. Improperly glazed ceramic articles can release toxic levels of lead into acidic foods such as fruit juices and tomatoes.

Notes J.C. Meranger of Health and Welfare Canada: "The health hazard associated with the leaching of lead from ceramic glazes by acidic juices is well documented and several cases of poisonings have been reported." One youngster in Montreal actually died as a result of a month's drinking of large quantities of apple juice that was being stored in a modern, handmade earthenware jug. The apple juice contained 157 parts per million lead after three hours' storage and *1300* ppm after three days!

Michael Klein and his colleagues, who investigated this poisoning case, found that 50 percent of all the glazed surfaces they tested were unsafe for table use (released over 7 ppm lead). These ceramics could give rise to chronic lead poisoning. Twenty-five percent of all domestic, handcrafted ceramics, and 10 percent of imported and commercial ceramics released over 100 ppm lead, the level expected to result in severe acute poisoning in small children.

The doctors in Montreal conclude: "Though the reported frequency of lead poisoning from pottery has been low, the true figure may be considerably higher. Fortunately, relatively few earthenware containers are used for storage of acidic solutions, and many are used only for decorative purposes. Public demand for hand-made pottery is leading to increased production and availability. Unless this is matched by an increased awareness of the problem by potters and governments, one can expect to see an increase in lead poisoning from this source."

Mercury

Mercury has been recognized as an environmental problem since the early 1940s, with food poisoning episodes occurring since then in Canada, the United States, Japan, Sweden, Pakistan, Iraq, and Guatemala.

There are about 3,000 uses for mercury in its various forms. The sources of mercury of greatest concern regarding air and water pollution, and hence food pollution, include the chlor-alkali industry (it no longer uses much mercury, but its past indiscretions are reflected in the mercury content of many waterways), fossil fuel combustion, and broken thermometers.

The mercury source in many past poisoning episodes was grain seed treated with methyl mercurial "dressings" as a fungicide. This use of mercury began in Canada in 1929, but was phased out entirely in the early 1970s (after wild bird poisonings in Alberta) and replaced with a diazinon-captan-lindane combination.

This switch is not entirely without its own drawbacks, since a U.S. Environmental Protection Agency study in 1980 indicated that lindane causes cancer, nerve damage, aplastic anemia, and birth defects, and recommended the pesticide be banned. Lindane and its breakdown products are commonly found in waterways in the Prairie provinces of Canada, due to the large amount of grain seed treated with lindane in that part of the country. By the fall of 1980, the Canadian federal government had made the decision not to ban lindane, calling the EPA study "unscientific" and "politically popular."

When talking of mercury's toxicity, it is important to distinguish between inorganic and organic forms of the metal, since most of the latter are much more poisonous and much more readily absorbed by living organisms. Mercury which appears in food is usually methyl mercury, an organic form. But pollution by inorganic mercury must not be ignored, since in 1967, Swedish scientists discovered that bacteria in the sediments of lakes and rivers could convert mercury from inorganic to organic forms, a process called bio-methylation. For this reason, *all* types of mercury pollution are highly dangerous. A gradient of mercury toxicity, from lowest to highest, is as follows: inorganic mercury salts and phenyl and methoxyethyl mercury salts (the latter two of which are organic), mercury vapour (inorganic), and methyl and ethyl mercury salts (both organic).

The primary threat posed by mercury-contaminated food is damage to the nervous system. This threat was elucidated all too clearly at Minamata, Japan, beginning in the 1950s; effects are still being felt. From 1932 on, effluent from the Chisso Corporation plant, an industrial chemical and fertilizer company, had contained mercury, which accumulated in fish in Minamata Bay and then in people eating the fish. Minamata disease is manifested in a myriad of symptoms: numbness, tunnel vision, hearing and speech impairment, loss of motor coordination sometimes leading to paralysis, birth defects, coma, and death.

Methyl mercury is extremely persistent. In the controversy over clean-up of mercury in the Wabigoon and English Rivers in north-western Ontario, estimates place the "life" of the mercury in the sediments at upwards of perhaps 70 years. Furthermore, while methyl mercury passes up the food chain, it bio-concentrates, because most organisms retain (rather than excrete) most of the methyl mercury they ingest. For example, tuna, a carnivorous fish which grows to be both large and old, can contain particularly high mercury (and lead) levels.

Mercury in any form is not easily taken up by plant roots, so levels of the metal in plant foods are very low. Ninety percent of any methyl mercury taken orally by humans is absorbed into body tissues, in contrast to the low absorption rates of cadmium and lead. Once absorbed, mercury heads especially for the liver, kidneys, and brain.

A very small proportion of the mercury found in food is naturally present, since mercury is widely distributed in nature. Levels of 0.1 to 0.5 parts per million in uncontaminated soils are not uncommon. Its natural occurrence in food is particularly evident in fish, because of bio-concentration. For example, in the Pinchi Lake mercury fault zone in British Columbia, fish routinely contain about 0.5 ppm mercury. (Much higher levels are found in fish from Pinchi Lake itself, of the order of 8 ppm, because of wastes from an abandoned mercury mine.)

Reports of mercury in plant foods are rare. Canadian examples include contamination of home garden vegetables near a chlor-alkali plant in Cornwall, Ontario and of mushrooms produced by several

Ontario growers (due, it is believed, to the horse manure component of the compost used, since several drugs used to treat horses contain mercury). Generally, though, plant foods rarely contain more than traces (e.g., 0.02 to 0.05 ppm) of mercury. Still, since methyl and ethyl mercury accumulate in the body, food residues that are minute individually can contribute to a build-up to toxic concentrations in humans, making it difficult to establish a safe daily dietary intake.

Mercury in food plants and even livestock products pales in comparison to levels found in fish. This is true all over the world. The first concrete evidence of mercury-contaminated fish in Canada was in Ontario's Lake St. Clair late in the 1960s. Dow Chemical in Sarnia and Wyandott Chemicals in Detroit together had been dumping over 80 kilograms (200 pounds) of mercury *per day* into the lake.

The Lake St. Clair discovery stimulated investigations in other waterways close to industrial mercury sources, such as the Interprovincial Cooperative and Reed Paper chlor-alkali plants at Saskatoon, Saskatchewan and Dryden, Ontario, respectively. Such research uncovered mercury-tainted fish all across Canada. The Canadians most affected by mercury-contaminated fish are native people, since many rely heavily on local fish as a food source. *Table 45* shows mercury levels in fish consumed by native Canadians.

Other aquatic food sources for native peoples are even higher in mercury. For example, the National Research Council of Canada's report on mercury indicates a level of 27.5 ppm in ringed seal liver in Franklin Territory, 12.1 ppm in common merganser (duck) breast muscle from Ontario's English River area, and 37.2 ppm in hooded seal liver from the St. Lawrence River in Quebec.

Marine food sources tend to be lower in mercury (see Atlantic data in *Table 9*), since oceans are further removed from polluting industries. Still, among those people clearly affected by mercury in marine fish are the Japanese, since their diet is so high in fish.

Part of the Ontario government's fish testing program (discussed in Chapter 9) consists of monitoring fish and warning sports anglers about mercury in lakes and rivers across the province. Mercury has shown up in areas remote from industrial sources, probably as a result of both natural contamination and aerial fallout of mercury pollutants. Mercury levels in the English-Wabigoon Rivers system have declined since Dryden's Reed Paper stopped using mercury in 1975, although they are still at unacceptable levels for continual consumers of fish

Table 45 / *Mercury Levels in Fish Consumed by Canadian Native Peoples*

Location (river system or district)	Type of Fish (and tissue tested)	Highest Recorded Average Mercury Level (parts per million)
Northwest Territories (Mackenzie)	walleye (lateral muscle)	1.19
Northwest Territories (Keewatin)	lake whitefish (lateral muscle)	0.18
British Columbia (Fraser R.)	lake trout (lateral muscle)	5.78
British Columbia (Okanagan R.)	rainbow trout (muscle)	0.29
Alberta (Churchill)	walleye (lateral muscle)	0.35
Alberta (Saskatchewan R.)	sauger (dorsal muscle)	1.3
Saskatchewan (N. Sask. R.)	white sucker (dorsal muscle)	4.3
Saskatchewan (S. Sask. R.)	northern pike (dorsal muscle)	9.1
Manitoba (Saskatchewan R.)	yellow perch (lateral muscle)	0.69
Manitoba (Winnipeg R.)	walleye (lateral muscle)	1.43
Ontario (Wabigoon R.)	northern pike (lateral muscle)	15.17
Ontario (English R.)	northern pike (lateral muscle)	4.75
Ontario (Winnipeg R.)	northern pike (lateral muscle)	1.38
Ontario (St. Lawrence R.)	walleye (lateral muscle)	2.88
Quebec (St. Lawrence R.)	northern pike (lateral muscle)	1.37
Quebec (Bell-Nottaway R.)	walleye (lateral muscle)	0.97
New Brunswick (Salmon R.)	yellow perch (lateral muscle)	1.05
New Brunswick (St. John R.)	striped bass (muscle)	2.13
Atlantic coast	cod	0.02–0.23*
	haddock	0.07–0.10*
	crab	0.06–0.15*
	swordfish	0.82–1.00*
	tuna	0.33–0.86*

* range

Source: National Research Council of Canada (Associate Committee on Scientific Criteria for Environmental Quality). *Effects of Mercury in the Canadian Environment.* NRCC Publication No. 16739. Ottawa: National Research Council of Canada, 1979.

from those waterways. A limited resumption of commercial fishing in Lake St. Clair was proposed by the Ontario Ministry of Natural Resources in 1980 (it was banned outright in 1970), although mercury levels in the fish there are dropping slowly.

Health and Welfare Canada's 1972 study of heavy metals in a *typical* Canadian diet (not high in local fish) found mercury intake to be about 10 micrograms per day. The Food and Agriculture Organization and the World Health Organization recommend a maximum level of 0.05 parts per million in food items other than fish. The 10 microgram figure translates to a concentration in food of less than 0.02 ppm, below the FAO/WHO guideline—but not much below.

Cadmium

Cadmium may be considered a "new" toxic threat. The reason is not so much that cadmium pollution is a recent phenomenon (it is not), but more because only recently have we discovered the extent of cadmium contamination and only recently have we appreciated the metal's high toxicity at low concentrations, relative to that of other metals.

Wherever there is zinc, there is cadmium; they occur together naturally in ore bodies, the cadmium being present at only 50 to 1000 parts per million in zinc. According to the National Research Council of Canada, the four major uses of cadmium in Canada are cadmium-plated metal, the manufacture of nickel-cadmium batteries (most of which are imported), pigments, and plastic stabilizers. Of particular interest regarding food contamination are lesser uses in medications for poultry and swine, and in some fungicides (mostly non-agricultural uses). Cadmium-containing pollutants, from metal smelting, pigment and alloy manufacture, the use of motor vehicles, and the use of super-phosphate fertilizers, find their way into soils and water supplies. As a result, sewage sludge intended for use as fertilizer on farmland often contains excessive levels of cadmium (see preceding section on sludge).

Unlike mercury, and luckily so, cadmium does not occur in organic forms; it's toxic enough in inorganic forms. It occurs naturally in soils, usually at concentrations less than 1 ppm. Unlike lead, which is tightly bound to soil particles, cadmium is easily transferred from soil to crop plants. It tends to remain in roots, so root crops (potatoes, carrots, beets, etc.) would be expected to have the highest cadmium levels. However, some leaf crops, especially lettuce, also accumulate cadmium in their leaves.

Technically speaking, cadmium does not bio-concentrate in the food chain. However, there is no rejection mechanism in the food chain to limit its upward passage; that is, there is no point in the chain where cadmium is not taken up. Furthermore, it does tend to accumulate in certain organs, especially kidney and liver. Therefore, cadmium's high availability and extreme toxicity warrant considerable attention with regard to food contamination.

The most notorious case of cadmium poisoning from food was "itai-itai" disease, which was a result of the consumption of contaminated rice in Japan beginning in 1935. Cadmium was not identified as the cause until 1961. This degenerative bone disease affected about 230 people. The probable source of contamination was mine tailings. Fortunately, only about 6 percent of any ingested cadmium is actually absorbed into body tissues (the rest is excreted), although diets low in calcium can allow absorption to rise to 10 percent.

The major source of cadmium intake by the general population in Canada is food, as opposed to air or drinking water. (Tobacco is a major cadmium source for smokers, since cadmium accumulates in tobacco leaves and, once inhaled, has no easy route out of the body.)

A typical daily dietary intake is of the order of 50 to 100 micrograms. The Food and Agriculture Organization and the World Health Organization recommend a maximum daily cadmium intake of 70 micrograms. Clearly, then, the levels of cadmium in our food are toxicologically significant. The cadmium ingested by "itai-itai" victims was only about 200 micrograms per day.

Cadmium levels in Canadian food in the period 1969 to 1971 are shown in *Table 46*. Communication in 1980 with one of the authors of the figures in the table, Diane Kirkpatrick of the Health Protection Branch of Health and Welfare Canada, indicated that current cadmium levels in our food are in the range of 0.01 to 0.05 parts per million.

No single food group is a proportionately large contributor to the total cadmium load. However, within food groups are some specific

Table 46 / *Cadmium Intake from the Diet in Canada, 1969 to 1971*

Food Group	Daily Consumption (grams per person)	Cadmium Range (parts per million)	Daily Cadmium Intake (micrograms per person)
Milk and dairy products	495	0.01–0.03	5–15
Meat, fish, and poultry	276	0.03–0.07	11–19
Cereals	190	0.04–0.07	8–13
Potatoes	191	0.06–0.10	12–19
Leafy vegetables	46	0.02–0.13	1–6
Legumes	32	0.04–0.05	1–2
Root vegetables	49	0.04–0.05	2–3
Garden fruits	82	0.02–0.04	2–3
Fruit	193	0.02–0.06	4–10
Oils and fats	27	0.05–0.10	1–2
Sugars and adjuncts	142	0.02–0.03	2–4
Drinks	59	0.02–0.03	1–2
Total			50–98*

* The Acceptable Daily Intake, according to the Food and Agriculture Organization/ World Health Organization, is 70 micrograms per person.

Sources: J.C. MERANGER and D.C. SMITH. "The Heavy Metal Content of a Typical Canadian Diet." *Canadian Journal of Public Health,* January–February, 1972, pp. 53–57.

D.C. KIRKPATRICK and D.E. COFFIN. "The Trace Metal Content of Representative Canadian Diets in 1970 and 1971." *Canadian Institute of Food Science and Technology Journal,* vol. 7 (1974), pp. 56–58.

_____. "The Trace Metal Content of a Representative Canadian Diet in 1972." *Canadian Journal of Public Health,* March–April, 1977, pp. 162–164.

culprits—see *Table 47.* Note that organ meats (liver, kidney) are cadmium-lovers. So are some seafoods, such as crab. Human breast milk is also worth watching, sometimes containing 35 to 95 parts per billion (0.035 to 0.095 ppm).

Total Metal Intake and "Safe" Levels

As part of its "total diet studies" looking at pesticide residues in foods,

Table 47 / *Cadmium Content of Specific Foods in Canada, 1971 to 1975*

Food	Cadmium Content (parts per million)
Apples	0.01
Cabbage	0.02
Carrots	0.03
Flour	0.04
Potatoes	0.04
Tomatoes	0.02
Milk	0.02
Eggs	0.03
Beef (muscle)	0.03
Pork (muscle)	0.03
Poultry (muscle)	0.02
Beef liver	0.15*
Pork liver	0.10*
Poultry liver	0.07*
Beef kidney	0.61*
Pork kidney	0.26*

* Concentration at questionably high levels

Source: D.E. COFFIN and W.P. MCKINLEY (Health and Welfare Canada). Unpublished data on chemical contaminants in the Canadian food supply, presented at the Fifth International Congress of Food Science and Technology, Kyoto, Japan, September, 1978.

the Health Protection Branch at Health and Welfare Canada has analyzed food samples from a typical Canadian diet for a number of heavy metals. Some of their results for 1969 and 1972 are shown in *Table 48.* These values do not give concentrations in the foods, but rather the total amounts in food based on a typical daily consumption of each food group. They do point out, for example, the relatively high contribution of the meat, fish, and poultry group (mostly the fish portion) to total mercury, while they obscure the relatively high nickel concentrations contributed by the oils and fats group, since we eat proportionately small amounts of those foods. (The nickel is probably a result of the use of nickel catalysts in the hydrogenation of vegetable oils.)

The authors of the 1969 Health and Welfare study conclude that "the relatively low levels of heavy metals found in the typical Canadian

Table 48 / Metals in the Canadian Total Diet, 1969 and 1972

Average Metal Contribution (micrograms per person per day)

Food Group	Cadmium '69	'72	Chromium '69	'72	Copper '69	'72	Lead '69	'72	Mercury '69	'72	Nickel '69*	'72	Zinc '69	'72
Milk and dairy products	15	7	55	74	84	95	10	18	5	0.4		43	2,571	2,593
Meat, fish and poultry	19	8	50	19	412	313	8	11	3	8.0		68	11,891	9,132
Cereals	13	7	29	27	531	430	13	8	4	0.3		63	2,800	1,907
Potatoes	19	10	50	24	484	225	4	6	2	0.3		34	1,295	806
Leafy vegetables	2	2	4	5	39	42	6	1		0.1		9	92	93
Legumes	1	<1	3	5	48	38	3	4	0	<0.1		16	168	245
Root vegetables	3	2	7	5	43	34	10	1	1	<0.1		11	106	99
Garden fruits	3	3	19	11	74	66	7	7	1	0.1		45	209	153
Fruits	4	3	14	59	100	105	73	14	2	0.1		29	168	167
Oils and fats	1	1	2	3	50	42		1	0	<0.1		40	370	174
Sugars	3	7	47	41	336	206	4	5	1	0.1		43	164	148
Drinks	2	1	4	10	16	10	1	1	1	<0.1		8	33	35
Total	85	51	284	283	2,217	1,606	1	77	21	9.4		409	19,867	15,552

* No data

< Less than

Sources: J.C. MERANGER and D.C. SMITH. "The Heavy Metal Content of a Typical Canadian Diet." *Canadian Journal of Public Health,* January–February, 1972, pp. 53–57.

D.C. KIRKPATRICK and D.E. COFFIN. "The Trace Metal Content of a Representative Canadian Diet in 1972." *Canadian Journal of Public Health,* March–April, 1977, pp. 162–164.

diet are an indication of the high quality of our food supply." This statement should be examined in light of the following concerns:

● Two metals of possible toxicological significance have not been included—arsenic (especially from livestock and marine food sources) and tin (from canned goods).

● The use of composites (i.e., food groups rather than individual food items), though a useful way of reporting results, does obscure relatively high metal contributions from individual foods. (A similar problem in reporting pesticide residues and deciding on pesticide tolerances was discussed in Chapter 5.) Also, because the diet is a "typical" one, the metal intake of someone eating "atypically" is not well represented. Consider, for example, an organ meat aficionado. Look at the figures in *Table 49*. If, out of the meat/fish/poultry composite, one eats only liver and kidney, one could be asking for trouble. The difference between metals in muscles versus organs is especially pronounced for cadmium.

Table 49 / *Lead, Cadmium, and Copper in Different Cattle Parts in the United States, 1971*

Metal	Part of Animal	Average Concentration* (parts per million)	Range of Values (parts per million)
Lead	Liver	0.54	0.01–3.74
	Muscle	0.36	0.01–2.96
	Kidney	0.63	0.02–4.90
Cadmium	Liver	0.21	0.01–3.17
	Muscle	0.08	0.01–1.00
	Kidney	0.55	0.01–7.82
Copper	Liver	34.92	1.30–123.00
	Muscle	1.82	0.03–29.90
	Kidney	4.18	0.08–24.20

* These figures are averages for positive samples, that is, those that contained detectable metal levels. Percent incidence of positive samples was 80 to 100 percent.

Source: W.A. RADER and J.E. SPAULDING. "Regulatory Aspects of Trace Elements in the Environment." *In:* Oehme, Frederick W. (editor). *Toxicity of Heavy Metals in the Environment.* New York: Marcel Dekker, Inc., 1978 (Part I) and 1979 (Part II).

● As noted for organic contaminants in fish in Chapter 9, using averages rather than maximum levels in reporting contaminant concentrations, while the norm in scientific work, does not really tell us how bad the poisoning can be. For that reason, it is most important that the *range* of concentrations found also be reported (as is done in *Table 49*).

● We must compare metal levels in food with those considered acceptable internationally. The information given in *Table 50* indicates that the picture is not exactly rosy, especially for cadmium and

---

Table 50 / *Canadian Daily Dietary Intakes of Cadmium, Lead and Mercury in Relation to Internationally Accepted Levels*

Metal	FAO/WHO Provisional Daily Tolerable Intake (micrograms per kilogram body weight)[1]	Canadian Daily Dietary Intakes (micrograms per kilogram body weight)			
		Adults[2]	Infants[3]		
			1 month	6 months	12 months
Cadmium	0.95–1.19	1.12	1.51	2.77	3.27
Lead	7.15	1.92	4.38	6.91	8.11
Mercury	0.72	0.20	—	—	—

[1] Calculated from the Food and Agriculture Organization/World Health Organization provisional tolerable weekly intakes.

[2] Based on Canadian total diet studies, 1969 to 1972.

[3] Based on Canadian baby food survey, 1975.

Source: D.E. COFFIN and W.P. MCKINLEY (Health and Welfare Canada). Unpublished data on chemical contaminants in the Canadian food supply, presented at the Fifth International Congress of Food Science and Technology, Kyoto, Japan, September 1978.

---

especially for babies. K.R. Mahaffey and his colleagues in the U.S. Food and Drug Administration (cadmium poses a similar, though actually slightly lesser problem in that country) state: "Cadmium is close enough to the tolerable intake so that further increases in the cadmium content of foods should be avoided."

Speaking of all heavy metals in the diet, the FDA scientists conclude that "it is generally considered that while the food supply contains less than tolerable intakes of those toxic heavy metals for which recommendations exist, any increases in these trace metal concentrations are undesirable."

Health and Welfare Canada has discontinued the total diet studies (see Chapter 5), but is continuing with a number of projects investigating heavy metals in foods. The health department is zeroing in on problem areas, such as lead in canned goods, metals in baby foods and mercury in fish.

Canadian Regulations on Metals in Food

Existing regulations and guidelines for metals in foods sold in Canada are shown in *Table 51*. These permissible levels apply only to what ends up in food; there are numerous regulations and guidelines at both the federal and provincial levels which state permissible metal levels in industrial wastewater effluents, in the air near pollution sources, and so on. For example: (a) Under the federal *Fisheries Act* (1977 Regulation), the amount of mercury that may be released by a chlor-alkali plant is 2.5 grams per day per tonne of chlorine produced; (b) Under the *Environmental Protection Act* of Ontario, the maximum amount of cadmium that can be emitted into the air by an industrial source is two micrograms per cubic metre of air per day and the maximum amount of lead is five micrograms per cubic metre of air per day.

The metal tolerances in *Table 51* have several shortcomings:

● There are *no* tolerances for cadmium in food, and discussions with Health and Welfare Canada officials indicate that none are forthcoming. The health department prefers to set tolerances only for metals and foods for which specific metal sources are consistently a problem; for example, lead in high-acid canned foods and tin in all canned goods. Cadmium, on the other hand, is present at low levels in almost all foods, because of indirect contamination through polluted air and water.

Without a specific tolerance, it would be difficult to condemn, say, beef kidney or liver for high cadmium levels because enforcement officials would have to fall back on Section 4 of the *Food and Drugs Act,* that nebulous section prohibiting a "poisonous or harmful substance" in food. No clear-cut, direct cause for high cadmium in these organ meats is evident and hence no easy way to reduce it is apparent,

Table 51 / *Levels of Metals Permitted in Canadian Foods*

(A) *Under the Food and Drugs Act*

Metal	Foods	Maximum Metal Content (Tolerance) (parts per million)
Lead	apple juice, cider, wine	0.5
	beverages as consumed and water in sealed containers other than natural mineral water or mineral water	0.2
	edible bone meal	10
	evaporated milk, condensed milk and concentrated infant formula	0.15
	fish protein	0.5
	fruit juice except apple juice	0.2
	ready-to-serve infant formula	0.08
Tin	canned foods	250
Arsenic	apple juice, cider, wine	0.2
	beverages as consumed and water in sealed containers other than natural mineral water or mineral water	0.1
	edible bone meal	1
	fish protein	3.5
	fruit juice except apple juice	0.1

(B) *Other Permitted Levels*

(1.) A guideline (not a regulation) has been set for total (inorganic and organic) mercury in commercial fish of 0.5 parts per million. This level is enforced by Fisheries and Environment Canada under the *Fish Inspection Act*, but only for fish products entering into interprovincial or international trade, including imports. For freshwater fish, application of the *Fish Inspection Act* has resulted in classifying lakes and rivers as open, restricted to certain species or sizes, or closed to commercial fishing according to mercury levels found in commercial fish. For marine fish, restrictions apply on a species and size basis and landings are monitored at registered plants.

(2.) A regulation under the *Hazardous Products Act* limits the concentration of lead in glazed ceramics to be used in storing, preparing or serving food to 7 parts per million and cadmium to 0.5 ppm. This regulation is enforced by Consumer and Corporate Affairs Canada, which monitors ceramics routinely for lead and cadmium.

Sources: Food and Drugs Act and Regulations (Table 1, p. 65A, Food and Drugs Regulations, Part B).
I. GRIFF SHERBIN (Environmental Protection Service, Environment Canada). *Mercury in the Canadian Environment.* (Economic and Technical Review Report EPS 3-EC-79-6.) Ottawa: Environment Canada, April, 1979.
Hazardous Products Act and Regulations (Glazed Ceramics Regulations).

so livestock producers might be unfairly treated by a regulation which would lead to confiscation of much of their beef liver and kidney. The elimination of the cadmium-based livestock medications wouldn't hurt.

● There are no tolerances for metals in fresh fruits and vegetables. Tolerances used to exist — 2 ppm arsenic and 7 ppm lead in fresh fruits, 1 ppm arsenic and 2 ppm lead in fresh vegetables — but were abolished in 1979. This was due in part to reduced use of lead arsenate as an agricultural pesticide. However, tolerances still exist for apples, the only crop where lead arsenate is still used (albeit to a very small extent) — 3.5 ppm lead and 1.0 ppm arsenic. (Note also in *Table 23* (Chapter 7), that residues of arsenic-containing livestock drugs are permitted in some meats.)

● The 250 ppm tolerance for tin in canned goods is probably much too high. Admittedly, this tolerance is widely accepted internationally, especially since absorption of tin into the body is slight. However, subtle toxic effects of tin are beginning to be recognized, such as effects on hemoglobin metabolism. In light of these findings, French toxicologist Claude Boudene recommends a reduction in the tin tolerance from 250 ppm to 100 ppm.

● There is reason to believe that enforcement of existing metal regulations under the *Food and Drugs Act* could be inadequate. An inspector with the Health Protection Branch, who preferred to remain anonymous, stated recently that the inspectors act only upon complaint, and generally rely on the food industry to monitor itself. For example, manufacturers of canned food check for lead and tin in their own goods.

It is encouraging to note that Canada has progressed further than the United States in regulating lead in food. While Canada has regulatory tolerances for lead in several foods (*Table 51*), the only U.S. limit is an "action level" of 0.5 ppm for lead in evaporated milk. Other proposed U.S. tolerances, which are higher than Canadian ones now in force, have met stiff opposition from the American canning industry.

Minimizing Metals

The most difficult sources of metals in food to reduce or eliminate are those that result from inadequate controls on waste disposal from metal-using industries and the ways in which they handle their waste products. We cannot afford another Love Canal or Wabigoon River; heavy industry must be required to install metal-retrieving pollution abatement technologies and research is required to determine safe methods of recycling and treating metals recovered in pollution abatement programs. Better monitoring of old industrial sites, abandoned mines, and water leaching from "sanitary" landfill sites is required, so that these metal sources can be contained.

University of Alberta geographer Rorke Bryan, in a 1973 survey of pollution sources in Canada, had this to say about industrial pollution abatement: "The mercury contamination problem calls into question the whole system of government control of the use of highly toxic chemicals for which no tolerance levels have been established, particularly in the absence of adequate supervision of their disposal. Obviously we can do little about the natural levels of mercury, but the mercury case shows clearly that we cannot assume that any form of pollution is safe until it has been proven harmful; rather, we must assume that all forms of pollution are harmful until they are proven safe. It should be incumbent on any government authority, firm or individual wishing to deposit waste products in the air, soil or water to demonstrate that no harmful side effects will result."

It should be added that the expert opinion of "non-vested" interests should be sought as well, to corroborate any claims of safety made by those wishing to donate their wastes to the environment.

One significant step forward is the replacement of lead in gasoline with other antiknock additives. An extra bonus in the introduction of "gasohol," that fossil-fuel-conserving mixture of 10 percent ethanol and 90 percent gasoline, is that the alcohol has antiknock properties.

All provinces of Canada should follow Ontario's example and develop programs for monitoring metals in sewage sludge, so that low-metal sludges can become a safe fertilizer resource for farmers and so that sludges high in one or more metals can be diverted from this use.

In August of 1979, the U.S. Food and Drug Administration announced that it wants to reduce lead intake from lead-soldered cans by at least 50 percent within the next five years. This is an eminently reasonable, if not rather conservative goal. At least four researchers of lead—Leon Russell, Jr., at Texas A & M University, Dorothy Settle and Clair Patterson at the California Institute of Technology, and Claude Boudene of the University of Paris-South—recommend the outright elimination of lead-soldered cans. Settle and Patterson conclude that lead-soldered tin cans contaminate their contents about tenfold over natural levels and that, since canned foods comprise about 20 percent of the U.S. (and, presumably, the Canadian) diet, elimination of lead-soldered cans would slash in half the dietary intake of lead.

Possible actions, then, to curb lead from cans would include, in decreasing order of effectiveness:

• outright elimination of lead-seamed cans—Aluminum is a metal of low toxicity, so lead-seamed cans could be replaced by aluminum cans (using recycled aluminum, since "virgin" aluminum consumes large amounts of energy in its manufacture); or, better still in terms of energy and resource conservation, recycleable glass jars (similar to those used in home canning).

• prohibiting the packaging in lead-seamed cans of *all* baby foods and *all* high-acid foods (tomatoes, fruit, fruit juices).

• better quality control in the use of lead solder in the canning industry, in addition to lacquering the interior surfaces of *all* cans to prevent tin migration into the food.

Efforts *have* been made in the can manufacturing industry over the past few years to reduce lead content of cans. A study by the Can Manufacturers Institute and the National Food Processors Association in the United States, reported in *Food Product Development,* found that such efforts have met with some success. The canned products surveyed in 1974 contained 0.32 parts per million lead on average, while the same products surveyed in 1976 contained only 0.19 ppm—a reduction of 41 percent.

What You Can Do

• Don't overindulge in organ meats—liver and kidney are good accumulators of metals, especially cadmium.

• Check on the extent of mercury contamination in a lake or river before fishing there. Restrict consumption of fish from heavily polluted waterways.

• Do not keep unopened canned goods for long periods, since lead levels in the food increase with time. One year should be the maximum storage period.

• Never leave a can containing food opened in the refrigerator, where exposure to oxygen will increase metal migration into the food. This goes for soups, fruit juices and so on. Transfer these foods to glass jars or plastic containers.

• Buy only juices that are packaged in glass, plastic, or paperboard containers.

• Don't cook high-acid foods (e.g., tomatoes, rhubarb) in metal or ceramic baking dishes. Use Pyrex instead.

• Make sure that all copper cooking pots, teapots, and so on have unblemished surfaces. They, and all other metal cooking surfaces, should be "seasoned" with oil to prevent corrosion.

• Any produce purchased at outdoor markets on busy city streets should be washed thoroughly, in order to remove surface residues of lead and other air pollutants.

11 A Potpourri

IT WOULD BE foolhardy to think that any single book could cover *all* the chemicals that *might* become unwanted guests at the dinner table. The preceding chapters have examined key groups of food contaminants. This chapter is a bit of a grab bag, a brief look at a few substances that don't fit neatly into any one of the previous contaminant categories. These are: contaminants from plastic packages; PCBs from paper packages; asbestos from filters used in the beverage industry; and radiation from nuclear power installations.

Plastic Packages

Section B.23.001 of Canada's *Food and Drugs Regulations* states the following: "No person shall sell any food in a package that may yield to its contents any substance that may be injurious to the health of the consumer of the food." We have already seen (in Chapter 10) that this Regulation has not stopped toxic levels of lead (from lead-soldered tin cans) from finding their way onto grocery store shelves. Similarly, plastic packages, though less of a toxicity problem than metal ones, are of some concern in specific instances.

The plastics industry has enjoyed spectacular growth in the last two decades. Roughly one-quarter of the plastics produced in North America are used as packaging materials, much of that for food packaging. Many food packages that were formerly made of paper, cardboard, metal, and glass are now made of some type of plastic, for many reasons—lower cost, lighter weight and increased durability are three.

But there are drawbacks to the use of plastics in food packaging. One is that since plastics can take so many forms and hence have so many different characteristics, they are very difficult to recycle and therefore become a serious garbage problem. One solution to the problem of recycling plastic containers is to refill them. The three-quart (2.4-litre), refillable plastic (polyethylene) milk jug used to capture a large percentage of the milk market, but retailers and dairies got tired of handling the empties. Now, Ontario is the only Canadian province where milk may still be purchased in refillable plastic jugs. A few U.S. states have retained them too.

The drawback to plastics in food packaging of concern in this book is their toxicity in certain instances. Basically, a plastic is formed by the reaction of many molecules of a simple compound (a monomer) to create a polymer or "macro-molecule." The most common plastic polymers used in food packaging are, in descending order, polyolefins (particularly polyethylene and polypropylene), polyvinyl chloride (PVC), and polystyrene (PS), most commonly found in the form of Styrofoam. There is experimentation in the food industry with other plastics such as polycarbonate for non-carbonated beverages and polyethylene terphthalate (PET) for carbonated beverages.

Added to the basic plastic are a number of chemicals (different ones for different plastics), which alter the physical characteristics of the plastic. Plasticizers make the plastic more flexible; stabilizers may ensure that the plastic does not deteriorate upon contact with oxygen, heat and light; and antioxidants such as the common food preservatives BHT and BHA (butylated hydroxytoluene and hydroxyanisole), used in plastic food packages that will contain fats, oils, and other foods susceptible to rancidity.

While in rare cases the plastic polymer itself is the toxic substance (usually they are chemically inert), most often blame lies with either the additives just mentioned or residues of unreacted monomer. While the toxicity of certain plastic components is well known when they are tested on their own (that is, not as a plastic component) in the laboratory, the key question is how "available" the toxic compounds are to the food once they are incorporated into the plastic. In other words, what ability does the toxin have to migrate from the plastic into the food? The following examples indicate some problems with migrating toxins.

● **Polyvinyl Chloride (PVC)**

The monomer for PVC production is vinyl chloride or VCM. Vinyl chloride is a known carcinogen, causing a rare form of liver cancer, angiosarcoma, in workers in PVC manufacturing plants. Some PVC contains unreacted VCM. It has been estimated by the International Agency for Research on Cancer that the uptake of vinyl chloride from food wrappings can be from 0.1 to 1 microgram per person per day—low, but measurable levels.

In 1976, the Canadian government placed restrictions on the use of certain PVCs in food packages. The effect of this move is demonstrated in *Table 52*—an 80 percent reduction in VCM. There has also been a move away from PVC for food packages generally, with the slack taken up by polyethylene, polypropylene, and polystyrene.

Table 52 / *Vinyl Chloride Content of Canadian Food Products, 1974 to 1976*

Product	Average Vinyl Chloride Content (parts per million)	
	1974	*1976*
Vegetable oils	2.16	0.42
Vinegar	1.93	0.09

Source: D.E. COFFIN and W.P. MCKINLEY (Health and Welfare Canada). Unpublished data on chemical contaminants in the Canadian food supply, presented at the Fifth International Congress of Food Science and Technology, Kyoto, Japan, September, 1978.

● **Polystyrene (PS)**

Two fingers may be pointed here, one at benzene, the feedstock (monomer) and another at the PS itself. As mentioned in Chapter 9, benzene is a carcinogen. It is an unavoidable contaminant in the present production of PS, which is used in its foamed (light-weight or Styrofoam) form for meat trays, take-out coffee cups, egg cartons, and the "bubbles" surrounding Big Macs, and in its non-foamed form for other plastics such as yogurt containers.

The Consumer Product Safety Commission in the United States is looking at benzene very closely, with a view to limiting the benzene content of PS food packages to 0.1 percent. Some packaging industry representatives argue that, with some additional expenditure in the manufacturing process, even trace amounts of benzene can be eliminated.

A curious problem with PS itself surfaced in 1979. Michael Phillips of the University of Connecticut Health Center discovered that tea to which lemon has been added, when placed in a PS cup, dissolves the PS. While drinking his lemon tea in a university cafeteria, Phillips saw the Styrofoam cup disintegrating before his eyes. He subjected the observation to laboratory analysis and concluded, in the *New England Journal of Medicine:* "Until proved otherwise, it seems likely that anyone who drinks lemon-flavored tea from a polystyrene container will also consume an appreciable amount of the container itself in solubilized form. Since polystyrene is carcinogenic in laboratory animals, I suggest that the time is now ripe to return to the tradition of drinking lemon tea from cups of fine bone china—a practice with more aesthetic appeal and less potential hazard."

Since Phillips' finding with lemon tea probably related to the acidic nature of lemon juice, a wise precaution might be to stay away from other acidic drinks in PS cups, such as tomato and apple juices.

● **Phthalic Acid Esters (PAEs)**
Phthalic acid esters are plasticizers. They impart flexibility to certain plastics, especially polyvinyl chloride. The acute toxicity of PAEs, when ingested by test animals, is low, but some researchers have reported teratogenic and mutagenic effects in the laboratory.

According to a 1977 Environment Canada report, the daily adult intake of PAEs from all sources (food, air, water, plastics used in medical practices, etc.) was estimated to be 100 to 10,000 times smaller than the internationally agreed upon Acceptable Daily Intake. But the report did note that the effects of prolonged exposure to low levels of PAEs are unknown and should be investigated.

Monitoring of plastic-packaged foods for PAE contamination is sporadic and data are virtually non-existent. Foods of the greatest concern are those with a high fat content, such as meat and dairy products. (Note that plastic milk bags do *not* contain PAEs; these bags are made of polyethylene.) Fresh produce is often wrapped in PAE-containing PVC, but the toxicity potential is low because produce is low in fat. However, food wraps used in the home often contain PAEs.

● **Acrylonitrile (AN)**

In the United States over the past five years, the polymer acrylonitrile has been on a merry-go-round of bans and ban cancellations on its use in plastic beverage containers. AN is a carcinogen. It is used in margarine tubs, vegetable oil bottles and plastic food wraps, but the focus has been on beverages, due to the large volumes consumed compared to the other foods mentioned. The controversy is over the extent (if any at all) of migration of AN into beverages.

There are no AN beverage bottles in Canada and little use of AN in food packaging generally. Food manufacturers are wary of using AN on the interior surfaces of any food packages.

In 1979, the federal Health Protection Branch conducted a survey of foods packaged in plastic. The study revealed the presence of acrylonitrile in some of the foods sampled. Analysis of the containers showed them to be fabricated from AN co-polymers harbouring residual AN. In February 1980, the U.S. Chemical Manufacturers Association issued the report of its two-year AN toxicity study. The study found that AN was carcinogenic in rats.

The Health Protection Branch presumes that AN could also be carcinogenic to humans. So in September 1980, the branch recommended that a regulation be added to the *Food and Drugs Regulations* prohibiting the sale of any food in a package that may yield *any* amount of AN to its contents.

Canadian regulations on the composition of plastic food packages—in fact, *all* food packages—are limited to the general proviso stated at the beginning of this section and the limitation on the use of certain PVCs. Unlike the United States, Canada has no lists of permitted or prohibited plastics and plastic ingredients for food packages. The U.S. regulations are based on the degree of migration of the plastic contaminant into the food. When migration levels are below a defined tolerance level, the plastic is not regulated. But if the plastic (or any other packaging material) is carcinogenic, and *any* level of migration is detectable, then the U.S. Food and Drug Administration usually restricts the plastic to non-food uses or to foods where migration would be almost nil because of the nature of the food (for example, PAEs in fresh produce wraps).

Instead of well-defined regulations on food packaging components, Health and Welfare Canada currently evaluates packages on an individual basis. If satisfied with a package material's safety, a "letter of no objection" is issued to the food manufacturer. The Health Protection Branch often uses FDA data and regulations as back-up documentation in coming to their own decisions on food packages.

Robert Ripley of Health and Welfare has noted that the health department's long-term goal is to develop a set of regulations on specific plastic materials, as the U.S. has done. Tolerances for plastic contaminants in food would make enforcement of the regulation prohibiting harmful packaging substances a lot easier.

PCBs in Food Packages

In July of 1971, the U.S. Food and Drug Administration's routine "market basket" analysis of food for pesticide residues uncovered relatively high levels of PCBs in shredded wheat. The contamination was traced to the packaging and then specifically to the cardboard dividers in the cereal box. Analysis of the cardboard revealed PCBs at levels of up to 433 parts per million.

Subsequent sleuthing uncovered similarly high PCB levels in crackers, bread crumbs, macaroni, pretzels, dessert mixes, milk powder, infant cereals (especially high levels), cookies, and other foods packaged in paperboard. The foods contained the common PCB "Aroclor 1242" at levels of up to 0.43 parts per million (average within a packaged food group), while the packages contained up to 54 ppm on average. Foods packaged in paperboard made from "virgin"—not recycled—materials had insignificantly low PCB concentrations.

How could this be? Another recycling idea gone sour: One component of paperboard packaging, which is comprised largely of reclaimed (recycled) paper products, used to be discarded "carbonless" copy paper, which contained 3 to 5 *percent* PCBs. Such paper was no longer being produced with PCBs in 1971. But, as a 1975 evaluation of the packaging problem by the FDA put it: "It became clear that unknown amounts of such PCB 'copy paper' on the nation's shelves, as well as the extreme stability of these compounds in the recycling process were causes for immediate concern for the safety of a large amount of the food supply packaged in paper."

As a result of the shredded wheat affair, the FDA slapped a maximum acceptable PCB level (tolerance) in food packages of 10 parts per million. Efforts by the paper recycling industry to exclude PCB-containing paper products from paperboard to be used in food packages has led to a decrease in PCBs in packaged foods, but continued strict quality control is essential since, by 1975, PCBs were still being found in packages at levels in excess of the 10 ppm tolerance.

A joint study in 1973 by Health and Welfare Canada and the Ontario Research Foundation found PCBs in Canadian food packages—see *Table 53*. The most contaminated products were refrigerated unbaked specialties, grated cheese and cheese products, dried fruits, and frozen fruit juices.

The Canadian researchers concluded that although PCB contamination of food packages does occur, the transfer of these residues to food is minimal. FDA researchers are more concerned. John R. Wessel, Scientific Coordinator of the Office of the Associate Commissioner for Compliance of the FDA has said: "Paper food-packaging material containing PCB's at 10 ppm or less can cause the food contents of the packaging to contain up to 0.6 ppm PCB's. This amount of PCB's represents 35 percent of the acceptable daily intake for PCB's. Indeed, paper packaging material is one of the primary sources of PCB's in the human diet. . . . There is no basis for eliminating or raising the tolerance for PCB's in paper food packaging material that is consistent with the demonstrated need to protect the public from unacceptable exposure to PCB's."

There is no legal tolerance or guideline for PCBs in food packages in Canada. PCB levels in recycled paperboard would be low or undetectable now, since PCB-containing carbonless copy paper stocks are depleted, but enforcement of a stated tolerance would be desirable nevertheless.

Asbestos

Well known for its fireproofing properties, asbestos, actually a group of complex inorganic substances, is valued in the food industry simply

Table 53 / *PCBs in Canadian Packages and Packaged Food*

(A) *Occurrence of PCBs in Packages*

Packaged Food Category	Percent of Package Samples Containing PCBs at:		
	Less than 1 ppm	1–10 ppm	More than 10 ppm
Crackers, bread crumbs, etc.	80.6	19.4	—
Macaroni and noodle products	95.7	4.3	—
Pretzels, chips, etc.	80.6	16.7	0.8
Breakfast cereals	87.1	12.9	—
Cake, biscuit, pancake mixes	77.0	19.7	2.3
Powdered milk	65.0	35.0	—
Dessert and pudding mixes	87.9	12.1	—
Infant cereals (dry)	65.7	34.3	—
Cookies	86.1	13.1	0.8
Refrigerated unbaked specialties	33.3	50.0	16.7
Rice, oatmeal, farina, etc.	81.0	16.7	2.4
Chocolate and cocoa products	70.9	18.2	10.9
Grated cheeses and cheese products	25.0	75.0	—
Dried fruits	51.9	29.6	18.5
Frozen fruit juices	26.3	10.5	63.2
Milk and milk products	100	—	—
Frozen vegetables	90.9	9.1	—
Potatoes	100	—	—
Frozen fish	96.6	—	3.4

(B) *Occurrence of PCBs in the Packaged Food Itself*

Food Samples Containing PCBs (parts per million)	Percent
less than 0.01	66.7
0.01–0.05	6.7
0.06–0.10	12.0
0.11–0.50	10.7
0.51–0.99	1.3
1–5	2.6

Source: DAVID C. VILLENEUVE et al. "Polychlorinated Biphenyls and Polychlorinated Terphenyls in Canadian Food Packaging Materials." *Journal of the Association of Official Analytical Chemists,* July, 1973, pp. 999-1001.

for its physical structure. It is composed of long, microscopically tiny fibres, which make an excellent filter in the processing of beer, wine, liquor, soft drinks, sugar, and lard. Unfortunately, the filtering process carries asbestos fibres from the filter into the food.

The health effects of *inhaling* asbestos fibres, which often occurs in occupational settings, are well known—asbestosis (severe scarring of the lungs) and lung cancer. However, the effects of *ingesting* asbestos have not been as thoroughly studied.

Much of the work of the toxicity of ingested asbestos has been done by two Health and Welfare Canada researchers, H.M. Cunningham and R.D. Pontefract. They have found that the small size of individual asbestos fibres allows them to penetrate the walls of the digestive tract and then to move freely throughout the mammalian body, lodge anywhere they like and sometimes form malignant tumors.

Asbestos may enter the food supply indirectly through the water used in food processing. Aerial fallout into waterways, erosion of rocks and soils, and the use of asbestos-cement (A/C) pipes for pumping drinking water, all contribute small amounts of asbestos. But the larger proportion of asbestos in food comes from the use of asbestos filters in the food industry. For example, the National Research Council of Canada's 1979 report on asbestos noted that municipal water supplies in Ontario contained in the order of 0.2 to 3.0 million fibres of asbestos per litre of water, while asbestos fibres in beverages included 11.7 million fibres per litre of Italian Vermouth, 6.6 million in Canadian beer, 2.5 million in orange soft drink, and 64.0 million in French wine.

In 1977, the Consumers' Association of Canada (CAC) studied the asbestos content of 15 best-selling red wines and found fibres in every case—see *Table 54*. On average, the Canadian wines were "cleaner," but not by much.

Armed with these test results, the CAC recommended banning the use of asbestos filters in *any* food processing. Cam Seccombe of the CAC noted that there are many other, less harmful kinds of filters available, such as diatomaceous earth or molecular cellulose filters, and that they are cost-competitive.

Since asbestos is a cumulative poison (that is, once ingested, we don't excrete it), its use in the processing of food and beverages sold in Canada should, indeed, be prohibited. No asbestos filters are allowed in the United States food industry; Canada must follow suit.

Radiation

Many chemical elements are radioactive. Their atoms can spontane-

Table 54 / *Asbestos in Red Wines Sold in Canada*

Label and Variety	Origin	Asbestos Content (fibres per litre)
Andres Moulin Rouge	Canada	427,000
Baptistin Caracous Cotes de Provence	France	1,100,000
Bertolli Chianti	Italy	640,000
Bright's Cresta Roja	Canada	1,280,000
Calona Sommet Rouge	Canada	768,000
Calvet Beaujolais Superieur	France	240,000
Chateau-Gai Marechal Foch	Canada	+
Damoy Granvillions	France	533,000
Franz von Metternich Nachf. Nierstein Kyffhauser	Germany	320,000
Jordan Wines Vallee Rouge	Canada	320,000
Prefontaines	France	1,170,000
Rioja Santiago Yago	Spain	747,000
Societe des alcools du Quebec Bordeaux	France*	64,000
Szekszardi Voros	Hungary	1,920,000
Vin Geloso Rouge Sec	Canada	427,000

* bottled in Quebec
+ below detection limit

Source: "Test: Asbestos in Wine." *Canadian Consumer*, June, 1977, pp. 44–46.

ously eject ("radiate") particles and energy waves from their nuclei. The three major forms of radiation which are given off in this process of radioactive decay are alpha and beta particles and gamma rays.

These forms of radiation harm living organisms by ionizing—that is, by altering the electrical charge of—the atoms and molecules in their cells. Ionizing radiation causes cancer and genetic effects at the cellular level, and so geneticists generally agree that there is no such thing as a safe dose of radiation.

Much of our exposure to ionizing radiation comes from natural sources—radioactive elements in the body, cosmic rays, and radioactive substances present in the natural environment. This "background" radiation exposure amounts to about 100 millirems per year (see 'rem' in the Glossary, Appendix B). But our radiation exposure from synthetic sources often equals or exceeds the background level. A large proportion of synthetic exposure is from medical and dental X rays. Other sources of radiation which have been "technologically enhanced"—that is, natural sources that humans have concentrated through industrial activities—include leakage of radioactive material from nuclear power facilities (uranium mines and mills, nuclear power stations) and fallout from atmospheric nuclear weapons testing.

Since low-level radiation—as opposed to Hiroshima-style high-level radiation—is ubiquitous, and harmful at any dose, every effort should be made to minimize "non-background" sources. These synthetic sources can cause radiation contamination of the food we eat.

● **Radiation Preservation of Food**
Yes! Radiation can be a deliberate food additive. Permission under Canada's *Food and Drugs Regulations* was first provided in the 1960s for the bombarding of onions and potatoes with gamma radiation from a Cobalt-60 source, to prevent sprouting. Gamma rays may also be used for deinfestation of flour. Just how radiation preserves food, whether it be for purposes of sterilization, pasteurization, disinfection, deinfestation, or sprout inhibition, is not precisely understood.

In the 1950s and 1960s, radiation preservation of food was hailed as a major breakthrough in food processing and a bright, profitable future for its use was envisaged. That never materialized. Notes B.H. Lauer of Health and Welfare's Health Protection Branch: "We have been in touch with officers of Radioisotope Division, Atomic Energy Control Board, who have advised us that there are no Cobalt-60 gamma irradiators in commercial use in the food industry at the present time."

Lack of a true need for the kinds of preservation that radiation offers (for example, potatoes and onions are often consumed long before sprouting would occur), the availability of other preservatives (for example, the chemical preservative maleic hydrazide for potatoes and onions), cost, and possible problems with consumer acceptance of irradiated food have slowed its adoption into common usage.

It is important to note that when foods *are* irradiated to preserve them, they do not come into contact with the radioactive source itself, but are simply exposed to gamma rays. Lauer of Health and Welfare emphasizes that *no* residue of any radioactive substance remains in the food. So don't worry about spuds that glow in the dark!

For further reading on radiation preservation of food, consult the source list for this chapter in Appendix C.

Microwave ovens can be a source of radiation for consumers, but only in the unlikely event that the oven is defective. In any case, microwaves, a form of non-ionizing radiation, do not become incorporated into food cooked in microwave ovens.

- **Environmental Sources of Radiation in Food**

Helen Caldicott is an Australian physician and antinuclear activist now practising pediatrics in Boston, Massachusetts. She's concerned about the effects of nuclear power on our food: "As a physician, I contend that nuclear technology threatens life on our planet with extinction. If present trends continue, the air we breathe, the food we eat, and the water we drink will soon be contaminated with enough radioactive pollutants to pose a potential health hazard far greater than any plague humanity has ever experienced. Unknowingly exposed to these radioactive poisons, some of us may be developing cancer right now."

Nuclear power plants and nuclear weapons are second only to medical/dental X rays as a radioactive hazard for humans. Radioactive isotopes of iodine, cobalt, krypton, cesium, plutonium, uranium, radon, and strontium are among the "hot" elements that appear in the food chain as a result of nuclear technology.

Norman Rubin, a nuclear energy and radiation researcher at Energy Probe in Toronto, maintains that the key radioactive substances to watch for as Canada's nuclear power program escalates are carbon-14 and tritium (hydrogen-3), since these elements are emitted in especially large quantities by Canadian (CANDU) reactors and since carbon and hydrogen form the basis of the molecules of living organisms. Also emitted as fission products from the routine operation of both Canadian and American nuclear reactors are iodine-131 and strontium-90.

Data on radiation levels in Canadian food are scarce. Three examples follow:

- As a result of nuclear weapons testing and the ability of debris from such testing to travel great distances before falling out, strontium-90 and cesium-137 have been found in the milk of Arctic mammals and Inuit (Eskimos) in northern Canada. The lichens and sedges that are the prevalent vegetation in much of the North retain a greater proportion of radioactive fallout than do other forms of vegetation. Since the Inuit rely on Arctic mammals as a food source and since radioactive substances tend to bio-concentrate, it is particularly high in these people. A study performed jointly by Macdonald College (of McGill University) and Health and Welfare Canada in 1967 found that Inuit milk contained 50 to 100 times more cesium-137 than samples of human milk taken in Montreal.

- Tritium has been detected at ten times the background level in tomatoes grown in greenhouses located 700 metres from the nuclear power plant at Pickering, Ontario. Ontario Hydro, operators of the power station, maintains that if a person eats 23 kilograms (50 pounds) of tomatoes a year from Pickering, he/she would be exposed to 0.1 millirem of radiation, whereas the maximum "safe" dose for a human for a year is way up at 500 millirems, and the typical Ontarian diet contains 30 millirems annually. So, as far as Hydro is concerned, the term "hot house" can be used only in the conventional sense.

- Ontario Hydro researchers have estimated that a person eating only food grown near the Pickering nuclear power plant would receive about 7.5 millirems per year from carbon-14 alone—50 percent higher than Canada's Atomic Energy Control Board has set as a "target dose" of C-14 from *all* sources.

- Low levels of radioactive contaminants from the 1979 nuclear accident at Three Mile Island, Pennsylvania are thought to have blown northward into eastern Canada. Possible food chain effects of the accident (near the plant) were documented in a 1980 investigative report in *Harrowsmith* magazine.

What You Can Do

- Stay away from beverages served in polystyrene (Styrofoam) cups, especially high-acid drinks like lemon tea, tomato juice and fruit juices.

- Press for the elimination of asbestos filters in the production of alcoholic beverages and soft drinks.

- Support stricter controls on the emission of radioactive wastes from nuclear power plants.

PART THREE

Summing Up

12 Where Do We Go From Here?

So that's the story there is to tell.

We have seen that our industrial society makes use of chemicals which are poisonous to humans and that one important route by which we may be exposed to these substances is the food we eat. Other routes of contamination are equally crucial to our overall health. All the clean food in the world would be for nought if our lungs are poisoned by polluted air and our drinking water is laced with toxic substances. Efforts to rid our diet of environmental contaminants must be made as part of the broader process of cleaning up the environment generally.

Many environmental food additives are persistent; they retain their toxic properties long after they have fulfilled their purpose. A good example is PCBs; they're terrific insulators in electrical equipment, but very harmful to living things once the PCBs are released into the environment through leakage or improper disposal. Such is their persistence that nursing mothers who have eaten PCB-laden food should seriously consider the toxicity of their milk to their youngsters.

The problem of PCBs in mothers' milk points up another issue I've tried to grapple with—weighing benefits and risks. Which is the overriding concern, the nutritional benefits of breast milk or the toxic risks of the chemicals that may concentrate in the milk? An easy answer eludes us here. Answers come a little more easily when good alternatives exist. For example, the benefits of heavy pesticide use touted by the agricultural establishment are far overrated, considering the benefits to human health of using the much lower levels of pesticides employed in the Integrated Pest Management (IPM) strategy for controlling farm insects.

As a society, our aim must always be toward zero risk to our health from chemicals in food, even though we'll probably never get there. When we're faced with a decision to use or not to use a toxic chemical, we must ask ourselves—do we really need it? And if we do, aren't there safer ways of doing the same thing? One could argue, for example, that we *do* need canned foods. But my chapter on metal pollutants clearly indicates that using lead solder is not a very safe way to seal the cans and that an alternative—non-leaded solder—does exist, though it is more expensive.

From the chapters on the key groups of environmental food additives (Part II), some concluding statements can be made:

• The large body of information that we have on the presence in food of PCBs, dioxin, mercury, and a few other toxic substances is the exception rather than the rule. The large amount of information on those chemicals has been a result of tremendous public pressure initiated either by an acute poisoning episode (such as the PBB fiasco in Michigan) and/or the work of a few dedicated (and often non-government) scientists. There has been adequate toxicological testing of only 1 to 2 percent of the roughly 300,000 synthetic chemicals used in modern industrial society.

• Most of the toxic chemicals that make their way into food through contamination of the general environment are far more serious threats in occupational settings, though involving far fewer people. Notable examples are pesticides, metals, and asbestos. But recognition of this fact must not preclude attempts to solve the two problems simultaneously. Who ever said you can't walk and chew gum at the same time?

• In the past decade, many steps have been taken to curb the release of toxic chemicals into the environment and, it follows, to curb the contamination of the food supply by these substances. Some contaminants are less of a concern now than in 1970, while others pose continuing problems—see *Table 55*. It would be safe to say that overall, our food contains lower levels of individual environmental additives now than it did 10 or 20 years ago. For example, DDT has been reduced to negligible levels in most domestic foods. As residues in food of individual contaminants decline, regulatory tolerances will

Table 55 / *Status of Environmental Food Contaminants: A Bird's Eye View*

Contaminant	Status
DDT and other organochlorine insecticides	Most uses banned in industrialized countries in early 1970s. Still in widespread use in less developed countries.
Pesticides generally	Total residues substantially lower now than 10 to 15 years ago. However, validity of current maximum residue limits (tolerances) is in doubt. Reduced pesticide use through Integrated Pest Management should lower residues further.
Occupational exposure to pesticides	Poisonings continue to be reported; many may remain unreported because of faulty diagnosis.
Livestock antibiotics	Initiatives taken by Health and Welfare Canada and in the U.S. by the Food and Drug Administration to curb further drug residues. However, a significant incidence of illegal residues still occurs. Resistance to antibiotics (in animals and humans) may be a bigger problem than residues.
DES (diethylstilbestrol)	Banned from use as a hormonal growth stimulant in cattle in 1974 in Canada and in 1979 in the U.S.
Ergot	Modern grain sorting and handling techniques prevent the ergotism that plagued societies prior to the twentieth century.
Aflatoxin	Low levels of aflatoxin continue to contaminate peanuts and peanut butter; the cold Canadian climate eliminates aflatoxin problems in domestic crops, but most peanuts are still imported.
PCBs (polychlorinated biphenyls)	New uses banned in Canada in 1979. However, accidents with old PCB-containing equipment and improper handling of PCB wastes mean that PCB residues in food and human milk are not yet declining.
Dioxin	Dioxin levels in herbicides much lower now than 20 to 30 years ago. No dioxin permitted in Canadian food. 2,4,5-T use restricted in some provinces and states. Dioxin showing up in several foods as more laboratories become capable of performing dioxin tests.
Lead	Continued use of lead-seamed tin cans contributes half the total dietary lead. However, a major environmental source—tetraethyl lead in gasoline—is diminishing.
Cadmium	No decrease in cadmium levels in food in recent years. No regulatory action (i.e., no tolerances). Better control on cadmium-containing industrial wastes required.

also be lowered, as has already occurred with respect to DDT over the past few years.

● Although *levels* of individual chemical residues may be down, the sheer *number* of different chemicals that creep into our diet is higher today than ever before, and there is no evidence to suggest that the overall total of contaminants is lower than it was 10 years ago. The problem posed by high residues of a few chemicals has gradually been replaced by the concern that the myriad of substances now invading the food supply may interact in unpredictable ways, possibly producing even more toxic combinations. The problem is as serious as it ever was.

There's plenty of work to be done to reduce or, where possible, eliminate environmental toxins in our food:

● Pollution—of air, water, land, food—is *not* the inevitable cost of using chemicals in agricultural and industrial technology, as some people might have us believe. Everything does *not* have to leak. There exist—even now—techniques to minimize the escape of toxic chemicals into our environment and their infiltration of the food supply.

● We will have to take action to clean up our food based on incomplete but reasonably predictive knowledge (from laboratory studies) of the ways in which specific chemicals affect human health. Such foreign substances must be presumed guilty until proven innocent. If we wait for the "ultimate" toxicological evidence on every toxic chemical, it will be too late to act to safeguard our health; the damage will have been done.

● We must ensure that government and the food industry (both agricultural and processing sectors) are constantly vigilant, continuously monitoring the levels of toxic chemicals in food. Canadian data on environmental food additives are sorely lacking in many areas. More research into local levels of contamination, not merely extrapolation of United States figures, is necessary.

● Permitted concentrations of toxic chemicals in food must be kept *well below* the Acceptable Daily Intakes established internationally by the Food and Agriculture Organization and the World Health Organization, primarily because of the interactions among individual toxins, often producing heightened toxicity, that are not usually accounted for when establishing "safe" levels for dietary intake.

● Regulatory (that is, legally enforceable) tolerances should be established for contaminants for which only administrative (non-enforceable) guidelines now exist.

● Not only do we need better regulation of environmental contaminants in our food, but we also need better enforcement of the regulations that already exist. A law is meaningless if the mechanism for ferreting out the offenders is faulty, let alone if there is little or no penalty for the transgression.

In some cases, curbing environmental food contaminants will mean higher price tags on food items. But if we take into account the potential broader social costs of eating toxin-laden food, especially the health care expenditures, the costs of eating toxin-free food are probably *much* lower in the final analysis. Consider the expenses of treating and educating a hyperactive child, whose condition may have been caused or aggravated by chemicals in food, especially artificial colour and lead. Consider the extra expenses, both monetary and emotional, of trying to cope with cancer rather than seeking to prevent it.

In the struggle for a clean environment—and hence clean food—the role of public interest organizations is crucial. Biologist and Nobel laureate George Wald put it this way when speaking about nuclear power: "Those of us who oppose nuclear power in its present forms have nothing to gain thereby but our share in the common good. Our opposition brings us into conflict with all the centers of power. It costs us our own money. It threatens rather than raises our professional status.... Whom is one to believe? One cannot be sure. But it helps to know that those opposed to nuclear power have nothing to gain from their position but the public good, that they are indeed willing to pay for the privilege of speaking out."

Public interest groups such as Pollution Probe are powerless unless they have a following. Please—use the recommendations peppered throughout this book, together with the names and addresses in Appendix D, to press those in positions of authority to make the necessary changes. It's in their own best interest; they eat too. Vote for candidates who demonstrate a sincere concern for environmental quality. In the meantime, while we're moving towards the clean food goal, remember the steps you yourself can take to minimize toxic chemicals in your food. These steps are summarized on the next page.

Waiting for Clean Food to Arrive

- Wash all fresh produce thoroughly, or, where possible, discard outer layers of the product, in order to remove some of the pesticide residues.

- Trim away excess fat from meat, and limit consumption of high-fat dairy products, since pesticides and some other organic contaminants concentrate in fatty foods.

- Grow your own food to the extent possible. Use ecologically based growing methods.

- For foods you cannot grow yourself, investigate the source of all products billed as organically grown.

- Individuals with severe allergies or other hypersensitive reaction to drugs used in livestock production should seek out "organic" (chemical-free) sources of animal products.

- New mothers who have reason to believe they have been unduly exposed to contaminated foods such as fish from polluted waterways, should have their breast milk tested for PCBs, DDT and metals, to determine whether nursing can proceed safely.

- Everyone should limit consumption of fish from polluted waters, especially large and/or carnivorous fish, which will have bio-concentrated any contaminants.

- Store all nuts, nut butters, grains and flours in a cool, dry place in sealed containers, in order to prevent the proliferation of toxic moulds (mycotoxins). Never swallow any shrivelled, discoloured, visibly mouldy, or off-taste nut.

- Limit consumption of organ meats—liver and kidney—due to possible heavy metal contamination.

- Do not store canned goods, especially high-acid foods like tomatoes and fruits, more than one year. Never leave an opened can in the refrigerator, but transfer the food to another container, in order to reduce lead contamination. Also, do not store fruit and tomato juices in ceramic containers.

- Do not buy any baby foods in metal cans. Also, try to buy all fruit juices in glass or cardboard containers.

- Do not bake high-acid foods in metal cookware. Use Pyrex instead. Also, make sure that copper cookware has not oxidized and keep all metal cookware well "seasoned."

Appendices

APPENDIX A

Toxins on the Job: Occupational Exposure to Pesticides

ATTENTION TO THE human health effects of pesticides must focus as much on occupational exposure as on residues in food. Pesticide dosages experienced by those who manufacture and apply these chemicals may be of the order of 100 to 1,000 times the levels found in food. As H.F. Kraybill of the U.S. Food and Drug Administration has put it: "Poisoning from pesticides can be expected to appear most quickly, most frequently, most diversely and most severely in those people most extensively exposed. A wide variety of biochemical, physiological and pathological aberrations have been observed from heavy exposure."

Furthermore, the means by which pesticides enter a worker's body are multiple; while residues in food are taken in only by ingestion, occupational exposure can occur by inhalation into the lungs and absorption through the skin, in addition to ingestion.

Time and again, the claim is made that if pesticides are manufactured and applied using all the recommended precautions—such as respirators, protective clothing, and stipulated intervals between spraying a field and reentering it—no poisonings will occur. Unfortunately, the time and expense involved in taking the necessary precautions make the temptation for abuse attractive.

The beauty (if one may call it that!) of the now-restricted organochlorine insecticides such as DDT was the low acute toxicity to workers and the relatively minor precautions required during application. Because of the threat of chronic health effects, the organochlorines were essentially phased out in the late 1960s and early 1970s, and replaced by organophosphate and carbamate insecticides, which have considerably higher acute toxicity. The effect of this change in insect control strategy on worker health began to emerge as early as 1968 when, for example, a significant increase in reported annual pesticide poisonings was noted in the cotton growing areas of the lower Rio Grande Valley in Texas.

The trade-off soon became apparent. We can use insecticides which are of low toxicity to farm workers but which persist as residues in food (i.e., the organochlorines) or we can apply chemicals which degrade quickly and hence do not pose a residue problem, but which because of this quick degradation, must be of high acute toxicity to insects—and, unintentionally, to humans (i.e., the organophosphates and carbamates).

Human exposure to pesticides during their manufacture and packaging can be more easily controlled than exposure on the farm, since relatively few workplaces are involved in the former. Farm exposure is, of course, much more dispersed geographically and involves a far greater number of people. Furthermore, it is controlled by persons who often have less understanding of the necessary precautions to be taken.

Health Effects of Occupational Pesticide Exposure

As reported in Wayland Hayes' *Toxicology of Pesticides,* records are available for workers who have been formulating DDT since 1945. These people remain well even while absorbing dosages equivalent to an oral intake of 35 milligrams per day. Study of liver function after 16 to 25 years of occupational exposure to DDT revealed no evidence of liver disease or abnormality of liver function. Examination of workers with more moderate exposure to DDT, but with exposure to a far wider range of pesticides, have also failed to reveal effects of medical significance.

Still, concludes Hayes, the effect of high occupational dosages of organochlorines in humans cannot be known with complete assurance until workers have spent their entire working lives in such employment. Kraybill reported that organochlorines *have* affected human health; workers exposed to these chemicals show characteristics of premature ageing and certain behavioural responses such as memory loss, inability to concentrate, and other neurological malfunctioning.

However, it is the dramatic effects of overexposure to the second-generation insecticides, the organophosphorous ones (such as parathion and malathion) and the carbamates (such as carbaryl and carbofuran), which have captured attention in the last decade. Symptoms of organophosphate and carbamate poisoning include nausea, vomiting, weakness, tremors, cramps, breathing difficulty, headaches, and involuntary urination and defecation. These symptoms are the result of the ability of these pesticides to decrease the activity of the nervous system enzyme cholinesterase. This enzyme is vital to the proper transmission of nerve impulses; depletion of cholinesterase will bring on poisoning. A depression of cholinesterase of 30 percent is considered a case of poisoning, whether or not there are overt symptoms.

In general, the organophosphates are more toxic than the carbamates, since, while cholinesterase inhibition caused by carbamates reverses itself when the person is no longer exposed, that caused by organophosphates does not reverse itself readily and may last weeks unless an antidote is administered.

Acute toxicities of common pesticides (expressed as the dosage required to kill half of test animals, the LD_{50}) were given in Table 4 (Chapter 5). (Note the very low LD_{50} of parathion compared to all other organophosphates and carbamates.) In addition to acute toxicity, some organophosphates and carbamates are at least suspected of having genetic effects such as carcinogenesis and teratogenesis.

Herbicides vary greatly in toxicity. Some, such as paraquat and dinoseb, are highly toxic and must be used with extreme caution. The ubiquitous 2,4-D is an irritant to skin, eyes, and the lining of the respiratory tract. It can cause muscular weakness, vomiting, diarrhea, and injury to the liver, kidneys, and nervous system. It is also highly suspected of being teratogenic.

Fungicides also vary widely in toxicity. The highly toxic mercury-based fungicides have all but disappeared from agricultural usage due to their health impacts. Highly toxic compounds such as DNOC and moderately toxic compounds such as dichlone and dodine are among those still used.

Synergy between pesticides is of particular concern to farm workers who are often spraying several pesticides in the course of one growing season. The most well known example is the interaction of malathion with EPN (o-ethyl-o-p-nitrophenyl benzene thiophosphonate). These two pesticides are each roughly five times more toxic (in terms of the LD_{50}) when used together as when each is used separately. Furthermore, a far greater cholinesterase depression results when the two pesticides are administered together.

The Global Picture

The World Health Organization estimates that there are some 500,000 cases of human poisoning by pesticides each year. On the farm worker front, these poisonings occur either when the pesticide is being mixed and sprayed, or when workers reenter treated fields to prune and harvest.

There is a large gap between the number of poisonings reported and the number that likely actually occurs, for three reasons. One is that farm workers may fear reprisals for reporting any effects of pesticides and may be unaware of compensation that is legally available to them in many countries. Another reason is that the effects may not be obviously attributable to pesticide exposure. A third reason may be that the jurisdiction (state/province, country) may not have in place a formal procedure for reporting pesticide poisonings.

A survey of farm workers in Tulare County, California in 1969 discovered that less than 1 percent of the pesticide illnesses serious enough to require treatment had been reported to workmen's compensation authorities.

Cases of pesticide poisoning abound:

● In Pakistan in 1976, some 2,500 people were poisoned by the organophosphate malathion, five of whom died as a result.

● Misapplication of pesticides and human poisoning is widespread in Latin American and other Third World countries. Illiteracy and language barriers mean that application instructions and precautions on package labels often go unheeded.

● One U.S. Food and Drug Administration official estimates, basing his figures on patterns of underreporting, that 80,000 persons in the U.S. are poisoned by pesticides each year and as many as 800 cases are fatal.

● The organophosphate insecticide leptophos (trade name: Phosvel), not registered for use in the U.S. but manufactured there for export to developing nations, was responsible for serious health effects among Velsicol Chemical Company workers in Bayport, Texas in the 1970s. Leptophos causes not only the cholinesterase inhibition typical of organophosphates, but also an unusual, sometimes crippling and often irreversible disease called delayed neuropathy, which may result in permanent paralysis some time after the victim has been exposed.

As reported in the journal *Environment* in 1977: "While Phosvel has never been cleared for use in the U.S., it has been marketed aggressively in other countries.... In Colombia, South America, for example, Phosvel is sold for use on rice, corn, sorghum, cotton, potatoes, tomatoes and beans. An advertising brochure from Velsicol Colombia describes Phosvel as being 1,470 times safer than parathion [re: cholinesterase depression].... There is no mention of Phosvel's neurotoxic properties.... To compare the two chemicals only on the basis of their respective ability to destroy cholinesterase is to make Phosvel appear unduly safe."

Poisonings in Canada

Statistics compiled by the Poison Control Program of Health and Welfare Canada's Health Protection Branch reveal that pesticides as a group are not a major cause of *reported* poisonings. However, without a mandatory reporting scheme for medically treated pesticide poisoning cases, we in Canada are in no position to sit back complacently and assume that no problem exists.

There are over 400,000 people in Canada employed in agricultural occupations. Professional exterminators handle only about 20 percent of agricultural pesticide application, so that 80 percent is in the hands of farmers and farm workers. Data on Canadian pesticide poisoning are sparse, but include the following:

● Farmers and produce packers in the Holland Marsh market gardening area north of Toronto have exhibited significant cholinesterase depression due to exposure to organophosphates such as parathion, malathion, and diazinon. However, the affected individuals do not show any abnormal clinical signs; that is, by outward appearance, poisoning is not evident.

● In 1980, an Alberta couple saw 17 doctors before discovering that weakness and headaches had been caused by dimethoate, a pesticide they had sprayed on an apple tree. Until dimethoate was detected as the cause, the couple's symptoms had been diagnosed as various psychological and physical conditions.

● In Saskatchewan in 1974, there were 83 documented cases of poisoning by the carbamate insecticide carbofuran, during a spraying program for grasshopper control. Some victims required hospitalization. Interesting was the poisoning of a five-year-old girl whose only contact with the insecticide was traced to her playing with a hat her father had worn during a carbofuran application.

● Although there are no reliable statistics on the health impact of working with the common herbicide 2,4-D, a general question in a survey of farmers and grain elevator operators in Saskatchewan, where vast acreages are sprayed, indicate some possible trends. As outlined in the National Research Council's report on phenoxy herbicides, 20 percent of the 3,330 workers surveyed responded that they had experienced ill effects from working with pesticides, 2,4-D apparently being the most troublesome. Symptoms were generally confined to the season or time of spraying and included nausea, loss of appetite, weight loss, and vomiting. A small number reported a skin rash. The annually recurring symptoms observed after several years were often more extreme than those observed in previous years, such that a number of farmers had to contract out their spraying to other operators. In others words, there is the suggestion of a possible development of sensitivity to 2,4-D with repeated exposure.

Controlling Occupational Exposure in Canada

Section 3 of the federal *Pest Control Products Act* states that "no person shall manufacture, store, display, or use any control product under unsafe conditions." Worker safety legislation not just for agriculture but for all occupations is the responsibility of provincial governments. Any such legislation should give the worker the right, without reprisal, to refuse to work under unsafe conditions.

The primary vehicles for informing farm workers about the hazards of pesticide use consist of the registration class of the pesticide and the instructions for precautions to be taken in application, given on the product label.

Regulations under the *Pest Control Products Act* divide registered pesticides into three use classes: Domestic, which can be used by anyone, including homeowners, and which, for this reason, are of relatively mild toxicity; Commercial, which may be used only in commercial operations such as agriculture, forestry, and industry and which are of greater toxicity than those in the Domestic class; and Restricted, which are commercial products for which severe limitations on use have been imposed because of relatively high toxicity.

All registered pest control products sold in Canada must state on the package, "READ THE LABEL BEFORE USING." Included on the label is a graphic poison symbol indicating caution, warning, or danger according to the degree of toxicity (danger being the slightest, caution the greatest). Similar graphics indicate flammability, explosiveness, and corrosion hazards. A precautionary statement on the label gives the measures to be taken by the applicator to prevent harmful effects on both himself/herself and the environment. Symptoms of poisoning, first aid instructions, and safe methods for disposal of empty containers must also be detailed.

A particularly important criterion for safe exposure to organophosphate and carbamate insecticides is the reentry interval, which is the required time period between last spraying of the crop and reentry to the field for picking. Only in some cases does the pesticide label give this information; in other cases, one must refer to provincial agriculture department crop production recommendations, which also reiterate other precautions to be taken when handling pesticides. (Examples for Ontario are given in *Table 56*.)

Reentry intervals are longer in jurisdictions such as California, where different weather and crop conditions require longer periods. For example, the reentry interval for parathion in many U.S. states is 14 days. Paraoxon, a highly toxic metabolite of parathion, has caused serious poisonings in California up to 45 days after spraying. There is considerable debate about the adequacy of current reentry intervals for protection of workers from cholinesterase depression.

All the precautions proffered on labels and in production recommendation handbooks have little effect if the worker chooses not to abide by them. This does happen. In a Saskatchewan study of carbo-

Table 56 / *Reentry Intervals for Pesticides Used on Ontario Vegetable Crops*

Pesticide	Reentry Interval (Hours)
Trade name/(common name)	
Lannate (methomyl)	24
Phosdrin (mevinphos)	24
Supracide (methidathion)	24
Birlane (chlorfenvinthos)	48
Dasanit (fensulfothion)	48
Furadan (carbofuran)	48
Guthion (azinphos-methyl)	48
Meta-Systox R (oxydemeton-methyl)	48
Monitor (methamidophos)	48
Parathion (parathion)	48
Phosvel (leptophos)	48
Systox (demeton)	48
Temik (aldicarb)	48

Source: Ontario Ministry of Agriculture and Food. *1979 Vegetable Production Recommendations.*

furan poisoning, reported by the National Research Council (NRC), all individuals exhibiting symptoms of carbofuran intoxication had not been following the recommended safety procedures for handling the insecticide, such as the use of respirators and gloves.

The NRC report notes: "Information on these procedures was distributed by the supplier in Saskatchewan and all individuals showing symptoms stated that although they were aware of the recommended safety precautions, the major reasons for not following the recommendations were complacency built up over the years while handling less toxic chemicals, anxiety or hurrying to get the spraying completed in the quickest and easiest way possible, and lack of availability of respirators, gloves. etc."

APPENDIX B
Glossary

Acceptable Daily Intake (A.D.I.)   An estimate of the amount of a food additive, pesticide residue, or other food contaminant that can be ingested without appreciable risk on a daily basis, over the course of a lifetime. The safety factors used in calculating A.D.I.s from animal No-Effect Levels (N.E.L.s) (arrived at through both acute and chronic toxicity studies) vary from 10 to 5,000, depending on the range of studies available, the toxic effects observed at higher dose levels, and several other factors. On a worldwide basis, A.D.I.s are recommended by the World Health Organization and the Food and Agriculture Organization. *See also* No-Effect Level; Theoretical Daily Intake

Acid Rain   *See* pH

Active Ingredient   The chemical compound in a pesticide product that is responsible for the toxic effects; that is, the toxicant. The active ingredient is combined with other, usually inert ingredients (sometimes called adjuvants) to produce a formulation which is in a suitable physical state for application. These adjuvants may consist of solvents, surfactants, carriers, etc. The active ingredient as manufactured commercially is called the Technical Material. *See also* Common Name; Formulation

Acute   Effects of a chemical that are immediate and short-term in nature. An allergic reaction is an acute effect of some food contaminants. An acute toxicity test refers to the effects produced in a short time by the test chemical when administered in a single dose. Acute effects are generally reversible effects; that is, the toxic effect disappears when exposure to the chemical ceases. Some chemicals can have both acute and chronic toxic effects. *See also* Chronic; Short-Term Toxicity Test

Adipose Tissue   Fatty tissue. Cells here contain large globules of lipids (fats). Because many organic pesticides and pollutants are fat-soluble, they tend to concentrate in adipose tissue.

Aflatoxins   *See* Mycotoxins

Ames Test   A quick and inexpensive means (relative to animal tests) of testing whether a chemical is a mutagen, and since most carcinogens are also mutagens, whether it is also a carcinogen. The Ames test uses specially developed strains of the bacterium *Salmonella typhimurium* and operates by differentiating enzyme activity between bacterial cultures containing or omitting the test chemical. Mutagenicity in bacteria has been shown to have a close (80 to 90 percent) but not perfect correlation with mutagenicity in mammals. *See also* Mutagen; Carcinogen

Antibacterials   *See* Antibiotics

Antibiotics   Drugs (also called antibacterials) that inactivate certain species that are toxic to other species. For example, several species of *Penicillium* fungi produce penicillin, which destroys certain bacteria. Antibiotics are often used in treating disease and promoting growth in livestock.

Background Radiation   *See* rem

Bio-Accumulation   *See* Bio-Concentration

Biocide   Any agent that kills a living thing. There is a fine difference in meaning between pesticide and biocide; biocide is a "killer of life," whereas pesticide is a "killer of pests." The term biocide takes into account the fact that forms of life other than target "pests" may also be killed by chemical pesticides and other toxic chemicals. *See also* Pesticide; Target Species

Bio-Concentration   The taking up of a chemical contaminant by an organism into its tissues—through ingestion, inhalation, or absorption through the skin—such that the amount present (either in the whole body or in specific tissues) is at a higher level than that found in the environment. Bio-concentration is a product of the ecological "food chain." Each organism eats many organisms from a step lower in the chain. The food organisms are digested, metabolized and excreted, but much of the chemical contaminant in them is retained. Thus, organisms accumulate residues and become contaminated, often from an environment that itself may appear relatively uncontaminated. Chemicals which bio-concentrate include the organochlorine pesticides, polychlorinated biphenyls, and some heavy metals. Other terms meaning essentially the same thing are bio-accumulation and bio-magnification. *See also* Concentration Factors

Biological Pest Control   In the narrow sense, the improvement of naturally operating predator-prey pest control by introducing new enemies or enhancing the effectiveness of existing enemies of the pest. In the broader sense, biological control includes the use of hormones and other biochemical attractants or repellants, and the release of sterile or otherwise genetically incompetent male insects. It may also include the breeding of pest-resistant plant varieties. Biological control is a cornerstone of ecological agriculture and a key component of Integrated Pest Management. *See* Ecological Agriculture; Integrated Pest Management

Bio-Magnification   *See* Bio-Concentration

Bio-Methylation   The transformation of inorganic mercury to methyl mercury (an organic form) by bacteria living in the sediments of waterways.

Bio-Transformation   The breakdown of a compound in a living organism to different compounds (metabolites). For example, humans can convert DDT to its metabolite DDE. The liver is the most important organ in the bio-transformation of foreign chemicals. *See also* Metabolite

Brand Name   *See* Trade Name

Cancer   An abnormal, uncontrolled growth of cells. A malignant tumour which spreads to destroy healthy tissues of the body. The ability to metastasize—to invade other tissues—is what differentiates cancerous from non-malignant (benign) tumours.

Carbamate Insecticides   A relatively new generation of insecticides. First marketed in 1958, they have come into common use only in recent years, as replacements for some of the organochlorine and organophosphorous insecticides. Also, they appear to be safer than the organophosphorous insecticides and are more selective in their toxicity to insects than either the organochlorine or organophosphorous insecticides. Like the latter, they act by inhibiting the enzyme cholinesterase in the target insect. Carbamates do not persist or bio-concentrate, which were two problems with the organochlorines. But their ecological and human health effects are not yet entirely understood; they are suspected teratogens in birds and mammals. The most commonly used carbamates are carbaryl (Sevin) and carbofuran (Furadan).

Carcinogen   Any agent—biological, chemical, radioactive—that causes cancer. (Also: carcinogenic (adjective).) A carcinogen is also considered to be a mutagen, but the reverse is not always true. *See also* Cancer; Mutagen

Chloracne   A skin condition caused by exposure to chlorinated hydrocarbon compounds, especially polychlorinated biphenyls (PCBs). *See also* Polychlorinated Biphenyls

Chlorinated Hydrocarbons   *See* Organochlorine Insecticides; Halogenated Hydrocarbons; Polychlorinated Biphenyls

Cholinesterase   An enzyme important to the proper conduction of nerve impulses. Organophosphorous and carbamate insecticides can depress the activity of cholinesterase. Applicators of such insecticides sometimes experience nervous system disturbances.

Chronic   Effects of a chemical that are long-lasting. Carcinogens cause chronic disease, both in the sense that cancer develops and spreads over a protracted period of time and in the sense that the onset of cancer occurs a considerable time after exposure to the offending chemical, often several years. A chronic (long-term) toxicity study refers to the effects produced by a test chemical when administered in repeated doses over a longer period of time, usually the major part of the expected life span of short-lived species, sometimes covering the entire life span and more than one generation of such species. Biological damage from chronic disease is irreversible. *See also* Acute; Short-Term Toxicity Test

Chronic (Long-Term) Toxicity Study   *See* Chronic

Clinical Ecology   A branch of medicine that involves the diagnosis and treatment of adverse physiological reactions caused by exposure to chemicals in the environment. The focus is preventive—the identification of causes, with their avoidance and/or treatment, as opposed to the more conventional medical approach of treating symptoms. Most people suffering from ecological illness have a family history of allergies and other symptoms. The most common irritants for susceptible individuals are foods, food additives, pesticides, other chemicals (gases, building materials, synthetic fabrics, etc.), particulate inhalants, stinging insects, and weeds.

Common Name (re: pesticides)   The simplified chemical name that is given to a pesticide's active ingredient by international agreement, regardless of who manufactures it. Common names are not capitalized. For examples of common names and trade names of pesticides, see *Table 56* in Appendix A. *See also* Active Ingredient; Trade Name

Concentration Factors (CFs)   Concentration factors (CFs) indicate whether or not a compound is accumulated in the tissue of an organism. The CF is calculated by dividing the concentration of the compound in the tissue (either on a whole body basis or in a specific organ such as the liver) by the concentration ingested in the diet or, with respect to fish and other aquatic organisms, taken up from the surrounding medium. A concentration factor less than or equal to one indicates that the compound does not accumulate in the tissues of the organism, while a factor greater than one indicates accumulation. *See also* Bio-Concentration

DDT   *See* Organochlorine Insecticides

Dioxins   A group of chlorinated hydrocarbons found as contaminants in pesticides such as 2,4,5-T and pentachlorophenol (PCP). The most toxic dioxin, and the one usually referred to, is tetrachlorodibenzodioxin (TCDD), which can be toxic to laboratory animals at the parts per trillion level. *See also* Phenoxy Herbicides

Dosage   The rate and concentration at which a toxic substance is administered to animals or humans. Dosage is a key concept in toxicology. The sixteenth century physician Paracelsus put it this way: *"Dosis sola facit venum"* or "Dosage alone determines poisoning."

Ecological Agriculture ("Organic" Agriculture)   A system of growing crops and raising livestock without the use of synthetic fertilizers and pesticides, relying, instead, on the use of naturally occurring materials and organisms, as well as proper crop rotation, for pest control, fertilization, and soil improvement, with the purpose of maintaining a stable, sustainable, balanced ecosystem. *See also* Integrated Pest Management

Ecological Illness   *See* Clinical Ecology

Ecotoxicology   The study of the effects of poisonous substances as they occur together with other toxins and other stresses in the environment. It contrasts with "classical" toxicology, which studies the effects of single chemicals on test animals under controlled, laboratory conditions. *See also* Toxicology

Entomology   The study of insects.

Environmental Additive   *See* Food Additive

Enzyme   A protein molecule that acts as a catalyst and therefore affects the rate of a chemical reaction in a living organism.

Fetotoxic   *See* Teratogen

Food Additive   Any chemical that becomes part of a food item—either deliberately or incidentally—that is not normally considered an ingredient of the food. Most processing additives are deliberate; for example, preservatives, colouring agents, flavouring agents. The environmental (incidental) additives include packaging materials that migrate into the food, pesticide residues, and air and water pollutants. Regulatory definitions for food additives differ between Canada and the United States. In Canada, only substances intentionally added to food during processing are included as food additives, while the U.S. definition is broader.

Food Chain   *See* Bio-Concentration

Formulation   The form in which a pesticide product is actually sold. The formulation consists of the active ingredient (the pesticide itself) and one or more adjuvants, inert ingredients which facilitate application. Two common formulations are wettable powders and emulsifiable concentrates. Others include dusts and granules. *See also* Trade Name; Active Ingredient

Fungicide   A chemical that kills a fungus, a common cause of many plant diseases. Fungicides are also often used as "seed treatments" to protect seeds from deterioration by fungi.

Half Life   An index of the rate of breakdown of a chemical, or its conversion to simpler (though not necessarily less toxic) compounds. The half-life concept is most relevant to highly persistent chemicals, such as the organochlorine insecticides.

Halogenated Hydrocarbons   Organic compounds in which one (or more) of the hydrogen atoms has been replaced by an atom of a group of atomic elements called halogens—fluorine, chlorine, bromine, iodine, and astatine. The two halogens figuring prominently in environmental food contamination are chlorine and bromine. *See also* Organochlorine Insecticides; Polychlorinated Biphenyls; Polybrominated Biphenyls

Halogens   *See* Halogenated Hydrocarbons

Harvest Interval   The time between the last pesticide application and the time when any pesticide residue in or on the crop has dissipated to a level that makes the edible portion of the crop safe to eat.

Hazard (Risk)   The probability that injury will result from the use of a chemical in a proposed quantity and manner.

Heavy Metals   Also referred to as trace metals or trace elements, because they normally occur in very low concentrations in the earth's crust, water, air, and living things. While some heavy metals are essential to life at low levels, all are toxic (and some are carcinogenic) at high concentrations. There are 38 heavy metals, but those which have received the most attention with respect to occupational and environmental hazards are mercury, cadmium, lead, arsenic, chromium, zinc, nickel, tin, copper, and selenium. *See also* pH

Herbicide   A chemical that kills plants, usually weeds. About six times as much money is spent on herbicides in Canada—for control of broadleaf weeds and annual grasses—as on insecticides. *See also* Phenoxy Herbicides

Hydrocarbon   *See* Organic

Incidental Additive   *See* Food Additive

Inorganic   Any chemical compound that does not contain carbon. *See also* Organic

Insecticide   A chemical or biological agent that kills an insect. The chemical insecticides, DDT for example, tend to be non-specific (that is, will kill or harm several other insects in addition to the "target" species), while the biological pesticides tend to be specific (for example, the bacterium *Bacillus thuringiensis* is effective against about 100 species of caterpillars, but is harmless to beneficial insects and people).

Integrated Pest Management (IPM)   A form of pest control that abandons chemical treatment as the first line of defence against pests, relying, instead, on biological, physical, genetic, and other ecosystem-based control programs, with heavy emphasis on monitoring pest populations to prevent "insurance-style" pesticide application. Though IPM does employ chemical pesticides as a last resort, it is still a big step toward non-chemical, ecological agriculture. *See also* Ecological Agriculture

IPM   *See* Integrated Pest Management

Itai-Itai Disease   "Ouch-ouch" disease (from the Japanese), caused by excessive intake of the heavy metal cadmium.

Larvicide   Any agent which kills an insect pest in the larval stage.

Lead   *See* Pica

Lipid   *See* Adipose Tissue

Maximum Residue Limit (M.R.L.)   *See* Tolerance

Metabolite   A breakdown product of an organic compound. For our purposes, a metabolite refers most often to an altered form of a pesticide. For example, DDT breaks down to DDE, parathion to paraoxon, and so on. There is no general rule regarding the toxicity of a metabolite relative to that of its "parent" compound. *See also* Bio-Transformation; Terminal Residue

Metals   *See* Heavy Metals

Minamata Disease   Poisoning due to excessive intake of the heavy metal mercury; named after Minamata Bay, Japan, where people were maimed and killed from consumption of mercury-contaminated fish.

Miticide   A chemical that kills mites (closely related to insects).

M.R.L.   *See* Tolerance

Mutagen   Any agent that causes a change (mutation) in the DNA (deoxyribonucleic acid, the genetic "information") of a cell's chromosomes. A mutagen may also be a carcinogen. *See also* Carcinogen

Mycology   The study of fungi (moulds).

Mycotoxins   Toxic chemical compounds produced by some species of mould under certain temperature, moisture, and storage conditions. They are persistent substances and may remain in a food product even after the moulds which produced them are killed. The most notorious group of mycotoxins are the aflatoxins, which may be found in some nuts and grains.

N.E.L.   *See* No-Effect Level

Nematicide   A chemical that kills nematodes, worms which infect some agricultural soils and parasitize underground plant parts.

Nitrofurans   A group of drugs used widely in the rearing of poultry and, to a lesser extent, swine. They constitute the only group of livestock medications now permitted that have been proven to be carcinogenic in laboratory studies. The most common nitrofurans are furazolidone and nitrofurazone.

No-Effect Level (N.E.L.)   The amount of a substance that can be included in the diet of laboratory animals without causing toxic effects. The N.E.L. is the basis for determination of the Acceptable Daily Intake. The A.D.I. is typically set at 1/100 of the N.E.L. *See also* Acceptable Daily Intake

Non-Target Species   *See* Target Species

Organic   In the strict chemical sense, any compound that contains carbon and hydrogen (or other elements substituted for hydrogen). An organic compound can also be called a hydrocarbon. These substances include both naturally occurring (proteins, carbohydrates, etc.) and synthetic (DDT, PCBs, etc.) compounds. Some 700,000 organic compounds are known to exist. But "organic" is also widely used to describe agriculture and gardening that use only fertilizers and pest controls that are derived from living things. The latter use of the term is best avoided because it is somewhat misleading; it can be replaced by the term "ecological agriculture." *See* Ecological Agriculture

Organochlorine Insecticides   A group of chlorinated organic compounds, typified by the king-pin, DDT, which came into widespread agricultural use after World War II. DDT was highly valued as an insecticide because it was cheap and effective. But mounting evidence of its ecological effects—bioconcentration and high persistence—caused many countries to ban most or all uses of DDT and many other organochlorines about 10 years ago. Some of the organochlorines (besides DDT, these include chlordane, heptachlor, dieldrin, lindane) may be carcinogenic and/or teratogenic. Relative to the organophosphorous insecticides, the acute toxicity of the organochlorines is low, but the organophosphorous compounds are not nearly as persistent. Most organochlorines are liver toxins.

Organohalide Compounds   *See* Halogenated Hydrocarbons

Organophosphorous Insecticides   A group of organic compounds which have been part of the "new wave" of insecticides replacing the much-maligned organochlorines. Rapid degradation is their chief advantage over the organochlorines, eliminating any great concern about residue build-up after chronic, low-dose exposure. But they have high acute toxicity in mammals, causing nervous system disturbances through effects on levels of the enzyme cholinesterase. The most commonly used organophosphorous insecticides include malathion, diazinon and parathion.

Parent Compound   *See* Metabolite

Parts Per Million/Billion/Trillion (ppm/ppb/ppt)   An indication of the concentration of a chemical in water, food, body tissues, etc., used when the level of the chemical is relatively low. A part per million is 1 part of a chemical to 1 million parts of food, by weight. To use a food-related analogy, 1 part per million is equal to 1 ounce of salt in 31 tons of potato chips, while a part per billion would consist of a pinch of salt in 10 tons of potato chips.

Pathogen   Any disease-producing micro-organism or material.

PBBs   *See* Polybrominated Biphenyls

PCBs   *See* Polychlorinated Biphenyls

PCP   *See* Dioxins

Pesticide   Any agent—chemical, biological, radioactive—that kills an animal or plant unwanted by man. Includes insecticides, herbicides, fungicides, nematicides, rodenticides, miticides. Sometimes "pesticide" is used to refer specifically to insecticides, when in fact the term is all-encompassing. *See also* Biocide

pH   The pH scale is used to measure the acidity or alkalinity of any aqueous (water-based) solution. (Technically, the pH of a solution is the negative logarithm of the hydrogen ion concentration.) The scale ranges from 0 (very acid) to 14 (very alkaline). A solution with a pH of 7 is neutral. The lower the pH the greater the acidity. Each unit on the pH scale represents a tenfold increase or decrease in acidity. For example, a solution with a pH of 4 is 10 times as acidic as a solution with a pH of 5. Foods with a low pH include tomato juice, lemon juice, and vinegar.

pH is an important measurement in any discussion of the biological effects of heavy metals, because metals become more toxic at lower pH. Acid rain, formed by the reaction of nitrogen and sulphur oxides (from coal-fired power plants, metal smelting plants, and motor vehicles) with rain, lowers the pH of lakes and rivers. *See also* Heavy Metals

Phenoxy Herbicides   A group of organic compounds which includes 2,4-D, 2,4,5-T, 2,4,5-TP and MCPA. 2,4-D is particularly valued in agriculture because it is highly selective (that is, it does not kill non-target plants—the crops!). But the phenoxy herbicides have shown themselves to be suspected teratogens in laboratory studies, and 2,4,5-T (not used in Canadian agriculture) is routinely contaminated with dioxins, especially TCDD, one of the most potent toxins known. Acute reactions to 2,4,-D are common among farmers, grain elevator workers, and victims of roadside and schoolyard spray drift. These herbicides are relatively non-persistent. *See also* Dioxins

Pica   Any abnormal appetite for non-food substances. A common example of pica is the consumption of flaking paint by children in decaying urban areas, where old paint may contain high lead levels. Lead poisoning in children is often traced to this source.

Plasticizers   *See* Polyvinyl Chloride

Pollutant   A natural substance occurring in unnatural concentrations as a result of human activities, or a synthetic substance occurring in an undesired location in any concentration.

Polybrominated Biphenyls (PBBs)   A group of organic compounds related chemically to polychlorinated biphenyls (PCBs). Their most common use is as a fire retardant additive in some plastics. They were banned from all uses in Canada in 1978.

Polychlorinated Biphenyls (PCBs)   A group of some 100 chlorinated organic chemicals, occurring in mixtures of several different compounds. Their stability makes them ideal fluids for many industrial uses, but this same persistence, coupled with poor controls on their handling in the past, has led to their widespread dispersion in the environment. PCBs are even less bio-degradable than DDT, to which they are chemically related. In fact, much of the blame for ecological effects shouldered by DDT may in fact have been more correctly shared with PCBs, so close is their chemical relationship. PCBs bio-concentrate, accumulating in human milk and adipose tissue. Health effects of PCB exposure may include reduced fertility, disruption of enzyme and immunity systems, and liver cancers. *See also* Chloracne

Polyvinyl Chloride (PVC)   A general term applied to the group of plastics formed by polymerization of vinyl chloride monomer (VCM). PVCs vary in properties such as flexibility and strength, depending on the additives (plasticizers) used. PVCs are commonly used as food-packaging materials.

Processing Additive   *See* Food Additive

Prophylactic   The use of drugs to prevent disease (re: drugs in livestock production).

PVC   *See* Polyvinyl Chloride

Radiation   *See* rem

Re-Entry Interval   The minimum interval of time (hours or days) between the application of a pesticide and the time when workers can safely reenter the fields. These periods may be either given on the pesticide label, or recommended by provincial agriculture departments.

rem (Roentgen Equivalent Man)   A measure of the amount of damage produced in a living organism by the radiation passing through it. "Background radiation," i.e., radiation from natural sources, is commonly about 100 millirems a year, or 1/10 of a rem. The annual per capita dose from medical and dental Xrays in North America is about 70 millirems.

Residue   *See* Terminal Residue

Risk   *See* Hazard

Rodenticide   Any agent that kills a rodent pest.

Roentgen Equivalent Man   *See* rem

Safety   The practical certainty that injury will not result from the proposed use of a chemical.

Short-Term Toxicity Test   Sometimes also called a subacute toxicity test, the short-term toxicity test refers to the study of the effects produced by a test chemical when administered in repeated doses over a period of up to 10 percent of the expected lifespan of the animal.

Subacute Toxicity Test   *See* Short-Term Toxicity Test

Subtherapeutic   Uses of drugs (in livestock production) for other than disease treatment; in other words, for disease prevention and growth promotion. *See also* Therapeutic

Sulfa Drugs (Sulphonamides)   Used primarily in rearing swine, in combination with penicillin and tetracycline for disease treatment, disease prevention, growth promotion, and feed efficiency. Also used to a lesser extent in rearing cattle and poultry. The most common sulfa drugs for livestock use are sulfaethoxypyridazine and sulfadimethoxine.

Sulfonamides   *See* Sulfa Drugs

Synergy   The capacity of two or more chemicals found together to have a greater total toxic effect than would be the effect of each chemical taken independently, then added together. That is, the whole is greater than the sum of the parts. Related terms: antagonistic: the total effect is less than the effect of the most active component alone; additive: the total effect is equal to the sum of the effects of the components taken independently; and independent: the total effect is equal to the effect of the most active compound alone.

Systemic   An application of a pesticide that enters the circulatory system of the plant or animal rather than attacking only the surface. Also, a pesticide residue that cannot be removed by peeling off the outermost tissues or by washing, as is the case with topical pesticide applications. If a pesticide kills by entering the digestive or circulatory system of an insect, rather than by surface action, it is said to be a systemic pesticide. *See also* Topical

Target Species   The species of insect, weed, fungus, etc. that a pesticide is intended to kill. Discussion of the environmental impact of pesticides usually focuses on their effects on non-target species. For example, DDT was especially toxic to wild birds, while organophosphorous and carbamate insecticides are especially toxic to bees. Humans are a non-target species for the toxic effects of many pesticides. *See also* Biocide; Pesticide, Insecticide

TCDD   *See* Dioxins

Technical Material   *See* Active Ingredient

Teratogen   Any agent—chemical, biological, radioactive—that causes a birth defect. A chemical that is teratogenic is also said to be fetotoxic.

Terminal Residue   A residue that persists in the environment and in food after the application of a pesticide. Simply put, it is what "ends up" in food. Terminal residues consist of the persistent components of the material originally applied plus any persistent metabolites which may arise. *See also* Metabolite

Theoretical Daily Intake (T.D.I.)   An estimate of the possible daily consumption of a pesticide residue assuming that all foods consumed contain the maximum permissible residue. The guiding "rule" is that the acceptable daily intake be kept higher than the T.D.I. The T.D.I., not as commonly encountered as the A.D.I., is sometimes used in circumstances in which it is difficult to determine actual pesticide intakes. *See also* Acceptable Daily Intake

Therapeutic   The use of drugs to treat disease (re: drugs in livestock production). *See also* Subtherapeutic

Tolerance (Maximum Residue Limit or M.R.L.)   The maximum concentration of a chemical substance (pesticide, antibiotic, heavy metal, etc.) that may be legally permitted ("tolerated") in food. Tolerances are usually expressed in parts per million (ppm) or in parts per billion (ppb). *See also* Parts Per Million/Billion/Trillion

Topical   An application of a pesticide which is made to the surface of the plant or animal. Also, a pesticide residue which is found only in the outermost tissues of a food crop (such as the peel) or right on the surface. If a pesticide kills through surface contact, for example through the skin of an insect rather than through ingestion, it is also said to have topical activity. *See also* Systemic

Toxic   Poisonous. A substance may exhibit acute toxicity in a certain species, or chronic toxicity, or both. *See also* Acute; Chronic

Toxicity   The capacity to produce injury. For a thorough discussion of the different kinds of toxicity, refer to *Toxicology of Pesticides* by Wayland Hayes, Jr. (see source list for Chapter 5.) Appendix C. *See also* Dosage

Toxicology   The study of toxic substances. *See also* Toxic; Dosage; Ecotoxicology

Toxin   A toxic substance. *See also* Toxic

Trace Elements   *See* Heavy Metals

Trace Metals   *See* Heavy Metals

Trade Name (re: pesticides)   The name that is given to a pesticide product by its manufacturer, distinguishing it as being produced or sold exclusively by that company. Trade names are capitalized. For example, Sevin is a trade name for carbaryl and Furadan is a trade name for carbofuran. *See also* Common Name

VCM   *See* Polyvinyl Chloride

Withdrawal Period   *See* Withholding Period

Withholding Period   The minimum time interval (hours, days, months) between last administration of a drug to livestock—through feed, water, injection, etc.—and time of slaughter or milking. The purpose of the withholding period is to permit drug residues to dissipate to zero levels in the carcass or milk.

APPENDIX C
Chapter Sources

Entries marked with asterisks indicate particularly good sources of information.

1 / Food Additives Revisited

DENNIS, HARRY. "Don't Mess With My Genes" (editorial). *Not Man Apart* (Friends of the Earth), September, 1980, p. 3.

* *The Food and Drugs Act and Regulations.* Supply and Services Canada. Under constant revision.

Health and Welfare Canada (Health Protection Branch). *Food Additives: What Do You Think?* (Report on Opinion Survey Conducted Summer 1979). Ottawa: Supply and Services, Canada 1980.

* ____. *Food Additives: What Do You Think?* (Highlights of an Opinion Survey, Summer 1979). Dispatch No. 48. Ottawa: Health and Welfare Canada, 1980.

____. *Food Additive Pocket Dictionary.* Ottawa: Minister of Supply and Services, 1980.

* PIM, LINDA R. (A Project of the Pollution Probe Foundation). *Additive Alert: A Guide to Food Additives for the Canadian Consumer.* Toronto: Doubleday Canada, 1979.

2 / Chasing Zero Risk

ALMEIDA, W.F., de MELLO, D., and RODRIGUES-PUGA, F. "Influence of the Nutritional Status on the Toxicity of Food Additives and Pesticides". *In* Galli, C.L., Paoletti, P., and Vettorazzi, G. (editors). *Chemical Toxicology of Food.* Amsterdam: Elsevier/North-Holland Biomedical Press, 1978.

BROWN, MARTIN. "An Orange is an Orange." *Environment,* July-August, 1975, pp. 6-11.

* CHANT, DONALD A. and HALL, ROSS H. "Ecotoxicity in Canada." *In Canadian Environmental Advisory Council: Annual Review, 1977-1978.* Ottawa, Minister of Supply and Services, 1979.

DICKSON, DAVID. "Chemicals in Food: New US Framework Proposed." *Nature,* 8 March, 1979, p. 110.

* EPSTEIN, SAMUEL S. *The Politics of Cancer.* Garden City, N.Y.: Anchor Press/Doubleday, 1979 (revised and expanded edition).

FISCHHOFF, BARUCH. "Informed Consent in Societal Risk-Benefit Decisions." *Technological Forecasting and Social Change,* vol. 13 (1979), pp. 347-357.

* FISCHHOFF, BARUCH, HOHENEMSER, CHRISTOPHER, KASPPERSON, ROGER E. and KATES, ROBERT W. "Handling Hazards." *Environment,* September, 1978, pp. 16-20, 32-37.

* FISCHHOFF, BARUCH, SLOVIC, PAUL, and LICHTENSTEIN, SARAH. "Weighing the Risks: Which Risks Are Acceptable?" *Environment,* May, 1979, pp. 17-20, 32-38.

GORI, GIO BATTA. "Diet and Nutrition in Cancer Causation." *In* Galli, C.L., Paoletti, P., and Vettorazzi, G. (editors). *Chemical Toxicology of Food.* Amsterdam: Elsevier/North-Holland Biomedical Press, 1978.

GRICE, HAROLD C. "The Acceptance of Risk-Benefit Decisions." *In* Galli, C.L., Paoletti, P., and Vettorazzi, G. (editors). *Chemical Toxicology of Food.* Amsterdam: Elsevier/North-Holland Biomedical Press, 1978.

JACOBSON, MICHAEL F. "Diet and Cancer." *Science,* 18 January, 1980, pp. 258-261.

* PAEHLKE, ROBERT. "Guilty Until Proven Innocent: Carcinogens in the Environment." *Nature Canada,* April-June, 1980, pp. 18-23.

PAGE, TALBOT. "A Generic View of Toxic Chemicals and Similar Risks." *Ecology Law Quarterly,* vol. 7, no. 2 (1978), pp. 207-244.

RAVETZ, J.R. "Public Perceptions of Acceptable Risks." *Science and Public Policy,* October, 1979, pp. 298-306.

* ____. "The Risk Equations: The Political Economy of Risk." *New Scientist,* 8 September, 1977, pp. 598-599.

SCHATZ, GERALD S. "A Closer Look at the Academy's Report on Food-Safety Policy." *News Report: National Academy of Sciences, National Academy of Engineering, Institute of Medicine, National Research Council* (U.S.A.), May, 1979, pp. 1, 4-5.

Science Action Coalition with Albert J. Fritsch. *Environmental Ethics: Choices for Concerned Citizens.* Garden City, N.Y.: Anchor Press/Doubleday, 1980.

* Science Council of Canada. *Canadian Food and Agriculture Sustainability and Self-Reliance: A Discussion Paper.* Prepared by Science Council Committee for the Study: Canada's Scientific and Technological Contribu-

tion to World Food Supply. Ottawa: Science Council of Canada, May, 1979.

* _____. *Policies and Poisons: The Containment of Long-Term Hazards to Human Health in the Environment and in the Workplace.* Report No. 28. Ottawa: Science Council of Canada, October, 1977.

"Science, Nonsense and Responsibility" (editorial). *Nature,* 23 August, 1979, p. 619.

STEINFELD, JESSE L. "Environment and Cancer: Detecting and Eradicating Hazards in Our Environment." *In Environment and Cancer* (A Collection of Papers Presented at the Twenty-Fourth Annual Symposium on Fundamental Cancer Research, 1971). Baltimore: Williams and Wilkins Company, 1972.

STEWART, WALTER. *Hard to Swallow (Why Food Prices Keep Rising and What Can be Done About It).* Toronto: Macmillan of Canada, 1974.

THOMASSON, W.A. "Dangerous to Your Health: Saccharin, Cancer, and the Delaney Clause." *Atlantic Monthly,* June, 1979, pp. 25-26.

* United States Washington, D.C. : U.S.E.P.A., Environmental Protection Agency. *This Rat Died In a Cancer Lab to Save Lives: Animal Tests Find Most Chemicals Aren't Killers.* January, 1980.

WELLFORD, HARRISON. "Behind the Meat Counter: The Fight Over DES." *Atlantic Monthly,* October, 1972, pp. 86-90.

WHELAN, ELIZABETH M. and SMITH, TERRENCE. "Of Mice and Men—and Risk in Foods." *Across the Board,* March, 1980, pp. 74-81.

ZWERDLING, DANIEL. "The Food Monsters: How They Gobble Up Each Other—and Us." *The Progressive,* March, 1980, pp. 16-27. (re: reasons for the high cost of food)

3 / A Catalogue of Food Risks

Malnutrition

BROAD, WILLIAM J. "New Strength in the Diet-Disease Link?" *Science,* 9 November, 1979, pp. 666-668.

* *Eating in America: Dietary Goals for the United States.* Report of the Select Committee on Nutrition and Human Needs, U.S. Senate (George McGovern, Chairman). Cambridge, Mass.: The MIT Press, 1977.

GORI, GIO BATTA. "Diet and Nutrition in Cancer Causation." *In* Galli, C.L., Paoletti, P., and Vettorazzi, G. (editors). *Chemical Toxicology of Food.* Amsterdam: Elsevier/North-Holland Biomedical Press, 1978.

HALL, ROSS HUME and HAAS, ANNE JONES. "Strategy for Wellness." *En-Trophy Institute Review,* November-December, 1979, pp. 1-14.

PIM, LINDA R. "Nutrition Notes." *Probe Post,* March-April, 1980, p.3.

* United States Department of Agriculture and United States Department of Health, Education and Welfare. *Nutrition and Your Health: Dietary Guide-*

lines for Americans. Washington, D.C. : U.S.D.A. and U.S.D.E.H.W., February, 1980.

Microbiological Contaminants

* Agriculture Canada. *Food Storage in the Home.* Publication 1695. 1979.
* Health and Welfare Canada. *Health Protection and Food Laws.* Revised 1979. ("Control of Pathogens and Mycotoxins in Food," pp. 25-29.)
* _____. *Food Safety: It's All In Your Hands.* Revised 1978.
_____. *Dispatch No. 32— Microbial Food Poisoning.* March 1974.
_____. *The Can With a Story.* Factsheet.
"Honey Linked to Infant Deaths." *Harrowsmith,* no. 15 (1978), p. 152.
IMMEN, WALLACE. "Watch That Turkey, 1 in 3 is Contaminated." *The Globe and Mail,* October 7, 1980.

Natural Toxins

ABELSON, PHILIP H. "Cancer—Opportunism and Opportunity." *Science,* 5 October, 1979, p. 11.
* BOYD, ELDON M. *Toxicity of Pure Foods.* Cleveland: CRC Press, 1973.
CAMPBELL, MONI. "Unhealthy Health Food: Let the Buyer Beware." *Probe Post,* July-August, 1979, p. 13.
* Food Protection Committee, Food and Nutrition Board, National Academy of Sciences—National Research Council. *Toxicants Occurring Naturally in Foods.* Washington: National Academy of Sciences—National Research Council, 1967. Publication 1354.
HOSKINS, F.H. "Naturally Occurring Toxicants from Food." *In* Galli, C.L., Paoletti, P., and Vettorazzi, G. (editors). *Chemical Toxicology of Food.* Amsterdam: Elsevier/North-Holland Biomedical Press, 1978.
LEE, C.Y. "Nitrogen Compounds in Vegetable Foods." *Environmental Contaminants in Foods.* Proceedings of the Sixth Annual Symposium, New York State Agricultural Experiment Station (Geneva). Cornell University, Special Report No. 9, November, 1972.
* LEWIS, WALTER H. and ELVIN-LEWIS, MEMORY P.F. *Medical Botany: Plants Affecting Man's Health.* New York: John Wiley and Sons, 1977.
* LIENER, IRVIN E. (editor). *Toxic Constituents of Plant Foodstuffs.* New York: Academic Press, 1969.
WHELAN, ELIZABETH M. and STARE, FREDERICK J. *Panic in the Pantry: Food Facts, Fads and Fallacies.* New York: Atheneum, 1977.

Miscellaneous

BARNETT, DONNA. "Dr. Morrison's Worry List: Why Canada's Assistant Deputy Minister of Health and Welfare has Trouble Sleeping at Night." *Harrowsmith,* no. 16 (1978), pp. 59-60.
PIM, LINDA R. "Unlisted Hazards." *Probe Post,* March-April, 1979, p. 4.

4 / Controlling Contaminants in Food

BENNETT, P.R. "Establishment of Residue Tolerances under Food and Drug Acts." Proceedings of the Twenty-First Annual Meeting, Agricultural Pesticide Society, Quebec City, August 6-8, 1974.

DELVILLE, ROBERT. "Reception of International Food Standards in National Legislation." *Food Drug Cosmetic Law Journal,* vol. 33, No. 6 (June 1978), pp. 292-304.

* ELIAS, P.S. "The Acceptable Daily Intake for Man (A.D.I.) as a Chronic Toxicity Index." *In* Galli, C.L., Paoletti, P., and Vettorazzi, G. (editors). *Chemical Toxicology of Food.* Amsterdam: Elsevier/North-Holland Biomedical Press, 1978.

* *The Food and Drugs Act and Regulations.* Supply and Services Canada.

GERARD, ALAIN. "International Food Standards and National Laws." *Food Drug Cosmetic Law Journal,* vol. 33, no. 6 (June 1978), pp. 281-291.

GRARTEN, DAVID. "Food Standards for the World Community are Within Reach." *Food Product Development,* vol. 13, no. 6 (July 1979), pp. 60-61.

HAIGH, R. "Harmonization of Legislation on Foodstuffs, Food Additives and Contaminants in the European Economic Community." *Journal of Food Technology,* vol. 13, no. 4 (August 1978), pp. 255-264.

Health and Welfare Canada. *Food Plant Inspection.* Dispatch No. 21. September 1972.

* _____ *Health Protection and Food Laws.* Revised 1979.

HUNTER, BEATRICE TRUM. *The Mirage of Safety: Food Additives and Federal Policy.* New York: Charles Scribner's Sons, 1975.

KIMBRELL, EDDIE F. "Food Composition Regulation and Codex Standards." *Food Drug Cosmetic Law Journal,* vol. 33, No. 3 (March 1978), pp. 145-150.

SMITH, D. JOE, JR. "Detention and Seizure of Imports by the Food and Drug Administration." *Food Drug Cosmetic Law Journal,* vol. 33, no. 12 (December 1978), pp. 726-733.

TENNEY, JANET. "Food Standards and the Consumer." *Food Drug Cosmetic Law Journal,* vol. 33, no. 3 (March 1978), pp. 151-160.

5 / Pesticides

Of Pests And Pesticides

Agriculture Canada. *Let's Talk About Pesticides* (Pamphlet). 1980.

* _____. *Pesticides: Their Implications for Agriculture.* Publication 1518. 1973.

Agriculture Canada (Food Production and Inspection Branch). *Pesticides— Data Handling Procedures.* Trade Memorandum, February 14, 1980.

* Agriculture Canada, Fisheries and Environment Canada, and Health and

Welfare Canada. *Pesticide Use and Control in Canada.* Prepared for: The Canadian Council of Resource and Environment Ministers Meeting of June 1-2, 1977. Ottawa: Minister of Supply and Services, February, 1978.

* BORAIKO, ALLEN A. "The Pesticide Dilemma." *National Geographic,* February, 1980, pp. 145-183.

* BRYAN, RORKE. *Much is Taken, Much Remains: Canadian Issues in Environmental Conservation.* North Scituate, Mass.: Duxbury Press, 1973.

CAINE, HARVEY M. "Pesticides and Pollination." *Environment,* November, 1977, pp. 28-33.

* CARSON, RACHEL. *Silent Spring.* Greenwich, Conn.: Fawcett Publications, Inc., 1962.

"Current Views on Pesticides" (editorial). *Canadian Medical Association Journal,* January 25, 1969, p. 222.

DAVIES, JILL. *Monitor in Cauliflower.* Canadian Broadcasting Corporation ("The Food Show"), November 11, 1979.

DOTTO, LYDIA. "Battling the Bugs." *Science Forum,* March-April, 1979, pp. 34-41.

_____. "The War on Weeds." *Science Forum,* March-April, 1979, p. 38.

"Environmental Health Research" (editorial). *Science,* 21 December, 1979, p. 21 (re: aminotriazole in cranberries)

* ESTRIN, DAVID and SWAIGEN, JOHN. *Environment on Trial: A Handbook of Ontario Environmental Law.* Revised and expanded edition edited by Mary Anne Carswell and John Swaigen. Toronto: Macmillan of Canada Ltd., 1978.

GARCIA, RICHARD and DAHLSTEN, DONALD. "The Price of Pesticides." *Not Man Apart* (Friends of the Earth), September, 1980, p. 16.

GIORNO, FRANK. "97 Pesticides' Tests Fudged." *Probe Post,* July-August 1980, p. 10.

"Herbicide Resistant Weeds." *Agri-book (Corn in Canada),* February, 1978, p. 66.

HOWARD, ROSS. "Ottawa Kept Pesticide Facts Secret for 3 Years." *The Toronto Star,* August 24, 1980.

LAVIGNE, YVES. "Pesticide Testing Figures Fudged by American Firm, Ottawa Says." *The Globe and Mail,* June 25, 1980.

MAYSE, SUSAN. "A Chemical Chronicle." *Environment Views* (Alberta Environment), October-November, 1979, pp. 14-16.

* McEWEN, F.L. and STEPHENSON, G.R. *The Use and Significance of Pesticides in the Environment.* Toronto: John Wiley and Sons, 1979.

NORMINTON, CLAUDIA. *Pest Control in Canada: A Review.* Ottawa: Canadian Federation of Agriculture, May, 1973.

* OELHAF, ROBERT. *Organic Agriculture: Economic and Ecological Comparisons With Conventional Methods.* New York: John Wiley and Sons, 1978.

* *The Pest Control Products Act and Regulations.* Supply and Services Canada.

Pollution Probe. *Phenoxy Herbicides: A High-Powered Battle Against Weeds.* Toronto: Pollution Probe, 1980.

Pollution Probe and The Canadian Environmental Law Assocation. "Ontario and Federal Governments Lax About Suspect Pesticides." (Brief to Ontario Pesticides Advisory Committee.) *Probe Post,* September-December 1980, pp. 10-11.

SNELSON, J.T. "The Need for and Principles of Pesticide Registration." *FAO Plant Protection Bulletin,* vol. 26 (1978), pp. 93-100.

SNIATYNSKI, GILLIAN. "Environmental Impacts of Pesticide Use." *Environment Views* (Alberta Environment), October-November, 1979, pp. 3-7.

SWITZER, CLAYTON M. *The Future of Canadian Agriculture.* 1979 Klinck Lecture. Guelph, Ont.: Ontario Agricultural College.

* VAN DEN BOSCH, ROBERT. *The Pesticide Conspiracy.* Garden City, N.Y.: Doubleday and Co. Inc., 1978.

VICARS, MARYHELEN. "New Weapons in the War on Pests." *Environment Views* (Alberta Environment), October-November, 1979, pp. 12-13.

* VON STACKELBERG, PETER. "Safety Last: Tests that Fail the Test." *Maclean's,* August 25, 1980, p. 49.

"Which Pest is Best?" (editorial). *Probe Post,* July-August, 1980, p. 6.

How a Pesticide Becomes a Residue

* FRANK, R., BRAUN, H.E., ISHIDA, K., and SUDA, P. "Persistent Organic and Inorganic Pesticide Residues in Orchard Soils and Vineyards of Southern Ontario." *Canadian Journal of Soil Science,* November, 1976, pp. 463-484.

GALSTON, ARTHUR W. "How Safe Should Safe Be?" *Natural History,* vol. 85 (April, 1976), pp. 32-35. (re: carcinogenicity of atrazine metabolites in corn)

* HAYES, WAYLAND, JR. *Toxicology of Pesticides.* Baltimore, Md.: The Williams and Wilkins Co., 1975.

HUMPHRIES, DONALD. "Residues of Farm Pesticides Found in Snow in Rockies." *The Globe and Mail,* February 8, 1979. (re: 2,4-D in the Prairies)

MILES, J.R.W., TU, C.M., and HARRIS, C.R. "Persistence of Eight Organophosphorous Insecticides in Sterile and Non-Sterile Mineral and Organic Soils." *Bulletin of Environmental Contamination and Toxicology,* vol. 22 (1979), pp. 312-318.

SOMERS, E. "Environmental Contaminants in Foods—Problems and Possible Solutions of the Seventies." *Environmental Contaminants in Foods.* Proceedings of Sixth Annual Symposium, New York State Agricultural Experiment Station (Geneva). Special Report No. 9. Cornell University, November, 1972.

STEWART, D.K.R., CHISOLM, D., and RAGAB, M.T.H. "Long Term Persistence of Parathion in Soil." *Nature*, vol. 229, (1971), p. 47.

Toxicity of Residues in Food

DECROSTA, ANTHONY. "Do Pesticides Cause Allergies?" *Organic Gardening*, January, 1979, pp. 93-97.

_____. "How Pesticides Cause Flu." *Organic Gardening*, February, 1979, pp. 118-125.

* _____. "The Pesticides in Your Body and What To Do About Them." *Organic Gardening*, April, 1979, pp. 136-144.

DUBOIS, KENNETH P. "Combined Effects of Pesticides." *Canadian Medical Association Journal*, January 25, 1969, pp. 173-179.

EISENBRAND, G., UNGERER, O., and PREUSSMANN, R. "The Reaction of Nitrite with Pesticides II: Formation, Chemical Properties and Carcinogenic Activity of the N-Nitroso Derivative of N-Methyl-1-Naphthyl Carbamate (Carbaryl)." *Food and Cosmetics Toxicology*, vol. 13 (1975), pp. 365-367.

* HAYES, WAYLAND, JR. *Toxicology of Pesticides*. Baltimore, Md.: The Williams and Wilkins Co., 1975.

* KRAYBILL, H.F. "Significance of Pesticide Residues in Foods in Relation to Total Environmental Stress." *Canadian Medical Association Journal*, January 25, 1969, pp. 204-215.

MACLENNAN, JOHN C. *The Ecologic Road to Good Health*. Toronto: Allergy Information Association, 1978.

"Pollutants May Be Cause of Decline in Male Fertility." *The Globe and Mail*, September 12, 1979.

RANDOLPH, THERON. *Human Ecology and Susceptibility to the Chemical Environment*. Springfield, Ill.: Charles C. Thomas, Publishers, 1962.

SMALL, BRUCE and SMALL, BARBARA. *Sunnyhill: The Health Story of the 80's*. Goodwood, Ont.: Small and Associates, Publishers, 1980. (re: ecological illness)

Policing Pesticides in Canadian Food

BUTTERFIELD, M.J. and MCKINLEY, W.P. (Health and Welfare Canada). *Pesticides, Health and the Consumer*. Paper presented at the University of Saskatchewan (Saskatoon), January 10, 1977.

* CHAPMAN, R.A. "Canadian Food and Drug Viewpoint of Pesticide Tolerances." *Canadian Medical Association Journal*, January 25, 1969, pp. 192-197.

COCHRANE, WILLIAM P. and WHITNEY, WAYNE. "The Canadian Check Sample Program on Pesticide Residue Analysis: Reliability and Performance." *In* Geissbuhler, H. (editor). *Advances in Pesticide Science, Part 3*. Oxford: Pergamon Press, 1979.

* *The Food and Drugs Act and Regulations*. Supply and Services Canada.

GOULD, ROBERT F. *Pesticides Identification at Residue Level*. Washington,

D.C.: American Chemical Society, 1971. (Advances in Chemistry Series 104.)

The Pesticide Residue Compensation Act and Regulations. Supply and Services Canada.

Observed Residues in Canadian Foods: A Bird's Eye View

* COFFIN, D.E. and MCKINLEY, W.P. "Chemical Contaminants of Foods." *In* Proceedings of the Fifth International Congress of Food Science and Technology (Kyoto, Japan, September 1978.) Amsterdam: Elsevier Scientific Publishing Co., 1979.

COFFIN, D.E. and MCKINLEY, W.P. (Health and Welfare Canada). Unpublished data on pesticide contaminants in the Canadian food supply, presented at the Fifth International Congress of Food Science and Technology, Kyoto, Japan, September 1978.

* DUGGAN, R.E. and WEATHERWAX, J.R. "Dietary Intake of Pesticide Chemicals." *Science,* vol. 157 (September 1967), pp. 1006-1010.

Food and Drug Administration (U.S.) "DDT Residues in Foods Decline." *FDA Consumer,* May, 1975.

Bio-Concentration: Pesticides in Meat and Milk

BOSHOFF, P.R. and PRETORIUS, V. "Determination of Toxaphene in Milk, Butter and Meat." *Bulletin of Environmental Contamination and Toxicology,* vol. 22 (1979), pp. 405-412.

FRANK, R. "Food Contamination With Pesticides." *Notes on Agriculture,* University of Guelph (Guelph, Ontario), vol. 8, no. 3 (December, 1972), pp. 15-19.

* FRANK, R., BRAUN, H.E., HOLDRINET, M., SIRONS, G. J. SMITH, E.H., and DIXON, D.W. "Organochlorine Insecticides and Industrial Pollutants in the Milk Supply of Southern Ontario, Canada—1977." *Journal of Food Protection,* vol. 42, no. 1 (January, 1979), pp. 31-37.

HAYES, W.J., DALE, W.E. and PIRKLE, C.I. "Evidence of Safety of Long-Term, High, Oral Doses of DDT for Man." *Archives of Environmental Health,* vol. 22 (1971), pp. 119-135.

"How Breast-Fed Infants Obtain Their DDT." *New Scientist,* August 26, 1976, p. 436.

* HOLDRINET, M. VAN HOVE, BRAUN, H.E., FRANK, R., STOPPS, G.J., SMOUT, M.S. and MCWADE, J.W. "Organochlorine Residues in Human Adipose Tissue and Milk from Ontario Residents, 1969-1974." *Canadian Journal of Public Health,* January-February, 1977, pp. 74-80.

* MES, JOS, CAMPBELL, DAVID S., ROBINSON, R. NEIL and DAVIES, DAVID J.A. "Polychlorinated Biphenyl and Organochlorine Pesticide Residues in Adipose Tissue of Canadians." *Bulletin of Environmental Contamination and Toxicology,* vol. 17, no. 2 (1977), pp. 196-203.

O'LEARY, J.A., DAVIES, J.E., EDMUNDSON, W.F., and REICH, G.A. "Transplacental Passage of Pesticides." *American Journal of Obstetrics and Gynecology,* vol. 107, no. 1 (1970), pp. 65-68.

Ontario Ministry of Agriculture and Food. *War on Warbles.* Factsheet. April, 1979.

* RITCEY, W.R., SAVARY, G. and MCCULLY, K.A. "Organochlorine Insecticide Residues in Human Adipose Tissue of Canadians." *Canadian Journal of Public Health,* vol. 64 (1973), pp. 380-386.

SASCHENBRECKER, P.W. "Levels of Terminal Pesticide Residues in Canadian Meat." *Canadian Veterinary Journal,* vol. 17, no. 6 (June, 1976), pp. 158-163.

SOMERS, E. "Pesticide Residues—Are They a Health Hazard?" *Chemistry in Canada,* September, 1969, pp. 21-23.

The Whole Ball of Wax: Total Diet Studies

DECROSTA, ANTHONY. "The Pesticides in the Food You Buy." *Organic Gardening,* March, 1979, pp. 124-138.

DENNIS, HARRY. "Danger: Imported Foods." *Not Man Apart* (Friends of the Earth), December, 1979, p. 17.

* GUNNER, S.W. and KIRKPATRICK, D.C. "Approaches for Estimating Human Intakes of Chemical Substances." *Canadian Institute of Food Science and Technology Journal,* January, 1979, pp. 27-31.

HARRIS, DAVID. "The Vegheads Fight Back." *Mother Jones,* June, 1980, p. 10. (re: California lawsuit)

* MCLEOD, H.A., SMITH, D.C., and BLUMAN, N. "Pesticide Residues in the Total Diet in Canada, V: 1976-1978." *Journal of Food Safety,* (1980), in press.

* SMITH, D.C. "Pesticide Residues in the Total Diet in Canada." *Pesticide Science,* vol. 2 (1971), pp. 92-95.

* SMITH, D.C., SANDI, E., and LEDUC, R. "Pesticide Residues in the Total Diet in Canada, II: 1970." *Pesticide Science,* vol. 3 (1972), pp. 207-210.

* SMITH, D.C., LEDUC, R., and CHARBONNEAU, C. "Pesticide Residues in the Total Diet in Canada, III: 1971." *Pesticide Science,* vol. 4 (1973), pp. 211-214.

* SMITH, D.C., LEDUC, R., and TREMBLAY, L. "Pesticide Residues in the Total Diet of Canada, IV: 1972 and 1973." *Pesticide Science,* vol. 6 (1975), pp. 75-82.

Pesticides in Third World Produce

ARNOW, LAURA. "Stemming the International Toxics Trade." *Not Man Apart* (Friends of the Earth), September, 1980, p. 15.

"California Sets Tougher Pesticide Regulations." *Not Man Apart* (Friends of the Earth), February, 1980, pp. 1-2.

* Food and Agriculture Organization/World Health Organization. *Pesticide Residues in Food.* Reports of the Annual Joint FAO/WHO Meetings. WHO Technical Report Series 458 (1970), 474 (1971), 502 (1972), 525 (1973), 545 (1974), 574 (1975), 592 (1976), 612 (1977), 615 (1978).

PAPWORTH, D.S. and PAHARIA, K.D. "Value of Pesticide Registration/Regulation to Developing Countries." *FAO Plant Protection Bulletin,* vol. 26 (1978), pp. 101-109.

PIETERS, A.J. "Internationally Acceptable Residue Limits and Their Implications for Registration Procedures." *FAO Plant Protection Bulletin,* vol. 26 (1978), pp. 120-122.

* WEIR, DAVID, SCHAPIRO, MARK, and JACOBS, TERRY. "The Boomerang Crime." *Mother Jones,* November, 1979, pp. 40-48.

Cutting Down on Residues

* ARCHER, T.E. "Stability of DDT in Foods and Feeds, Transformation in Cooking and Food Processing, Removal During Food and Feed Processing." *Residue Reviews,* vol. 61 (1976), pp. 29-36.

Environmental Protection Agency (U.S.). "Pesticide Programs: Rebuttable Presumption Against Registration and Continued Registration of Pesticide Products Containing Ethylenebisdithiocarbamates." *U.S. Federal Register,* vol. 42, no. 154 (August 10, 1977), pp. 40618-40627.

RIPLEY, BRIAN D. "Residues of Ethylenebisdithiocarbamates on Field-Treated Fruits and Vegetables." *Bulletin of Environmental Contamination and Toxicology,* vol. 22 (1979), pp. 182-189.

RIPLEY, BRIAN D., COX, DIANE F., WIEBE, JOHN, and FRANK, RICHARD. "Residues of Dikar and Ethylenethiourea in Treated Grapes and Commercial Grape Products." *Agricultural and Food Chemistry,* vol. 26, no. 1 (January-February, 1978), pp. 134-136.

* STREET, JOSEPH C. "Methods of Removal of Pesticide Residues." *Canadian Medical Association Journal,* January 25, 1969, pp. 154-160.

6 / Alternatives to Pesticides: Ecological Agriculture

Agriculture Canada. *Let's Talk About Pesticides.* (Pamphlet). 1980.

_____. *Pesticides: Their Implications for Agriculture.* Publication 1518. 1973.

Agriculture Canada (Hagley, E.A.C., Trottier, R., Herne, D.H.C., Hikichi, A. and Maitland, A.). *Pest Management in Ontario Apple Orchards.* Research Branch, 1978.

Agriculture Canada (McClanahan, R.J.) *Integrated Control of the Greenhouse Whitefly.* Publication 1469. 1972.

ASELAGE, JOHN M. "Effects of Pesticides on the Soil Microbiota." *Bio-Dynamics,* no. 132 (Fall 1979), pp. 11-18.

BENNETT, JENNIFER. "The Patented Seed: Another Step Toward Corporate Agriculture?" *Harrowsmith,* no. 22 (September, 1979), pp. 46-55, 84-86.

———. "A New Threat to Our Food Supply?" *The United Church Observer,* April, 1980, pp. 23-25.

"Biological Control of Plant Pathogens." *The IPM Practitioner,* vol. 1, no. 9 (September, 1979), p. 2.

* BOTTRELL, DALE. R. (Council on Environmental Quality). *Integrated Pest Management.* Washington, D.C.: U.S. Government Printing Office, December, 1979.

BROWN, MARTIN. "An Orange is an Orange." *Environment,* vol. 17, no. 5 (July-August, 1975), pp. 6-11.

Canadian Broadcasting Corporation (Radio). *Organic Agriculture.* Documentary. September 3, 1979.

Canadian Organic Growers. *Organic Growing: The Necessary and Viable Alternative for Supplying Canadians with Food.* Submission to the People's Food Commission, Toronto, Ont., February 25, 1979.

FILLIP, JANICE. "American Farmers and USDA Start to Take Organic Seriously." *Not Man Apart* (Friends of the Earth), September, 1980, pp. 18-19.

FOWLER, CARY. "Plant Patenting: Sowing the Seeds of Destruction." *The CoEvolution Quarterly,* Winter 1979-1980, pp. 34-38.

GOLDMAN, M.C. "Organic Farmers Win in Court." *Organic Gardening and Farming,* January, 1976, pp. 146-149.

———. "Strong Trends in Marketing Organic Foods." *Organic Gardening and Farming,* August, 1977, pp. 147-154.

* GOLDSTEIN, JEROME. *The Least is Best Pesticide Strategy: A Guide to Putting Integrated Pest Management into Action.* Emmaus, Penn.: The JG Press, 1978.

HARWOOD, RICHARD R. "Natural Weed Controls are Looking Good." *The New Farm,* September-October, 1979, pp. 56-58.

HAY, JAMES R. "Weed Control Without Chemicals." In: *Practical Alternatives to Chemicals in Agriculture.* Proceedings of a Conference held at Echo Valley Centre, Fort Qu'Appelle, Saskatchewan, October 30-31, 1978. Regina, Sask: University of Regina (Extension), 1979.

HILL, STUART B. "Agricultural Chemicals and the Soil." *In Chemicals and Agriculture: Problems and Alternatives.* Proceedings of a Seminar held at Echo Valley Centre, Fort Qu'Appelle, Saskatchewan, November 3-4, 1977. Regina, Sask: Canadian Plains Research Center (University of Regina), 1978.

———. "Biological Approaches to Pest Control." In *Proceedings of the P.E.I. Conference on Ecological Agriculture.* Charlottetown, P.E.I.: The Institute of Man and Resources/The Ark Project, 1979.

* ____. "Eco-Agriculture: The Way Ahead?" *Agrologist,* vol. 8 no. 4 (Fall 1979), pp. 9-11.

____. *Soil, Food, Health and Holism: The Search for Sustainable Nourishment.* Macdonald College (McGill University), Ste-Anne-de-Bellevue, Quebec, 1980.

HUFFAKER, C.B. and MESSANGER, P.S. (editors). *Theory and Practice of Biological Control.* New York: Academic Press, 1976.

* "It's Natural! It's Organic! Or Is It?" *Consumer Reports,* July, 1980, pp. 410-415.

KLEPPER, ROBERT; LOCKERETZ, WILLIAM; COMMONER, BARRY; GERTLER, MICHAEL; FAST, SARAH; O'LEARY, DANIEL; and BLOBAUM, ROGER. "Economic Performance and Energy Intensiveness of Organic and Conventional Farms in the Corn Belt: A Preliminary Comparison." *American Journal of Agricultural Economics,* vol. 59 (February, 1977), pp. 1-12.

LAIRD, ELMER. "Cooperation Between Farmers and Researchers." *In Practical Alternatives to Chemicals in Agriculture.* Proceedings of a Conference held at Echo Valley Centre, Fort Qu'Appelle, Saskatchewan, October 30-31, 1978. Regina, Sask.: University of Regina (Extension), 1979.

* LOCKERETZ, WILLIAM. "Can We Take the Chemicals Out of the Corn Belt?" *Horticulture,* September, 1977, pp. 14-16.

LOGSDON, GENE. "The Case For An Ecologically Clean Apple Orchard." *The New Farm,* September-October, 1979, pp. 44-49.

MAYALL, SAM. "Ecological Agriculture in the United Kingdom: Practical Experiences." In *Proceedings of the P.E.I. Conference on Ecological Agriculture.* Charlottetown, P.E.I.: The Institute of Man and Resources/The Ark Project, 1979.

MOONEY, P.R. *Seeds of the Earth: A Private or Public Resource?* Ottawa: Inter Pares (for the Canadian Council for International Co-operation and the International Coalition for Development Action, London), 1979.

* National Film Board (Canada). *A Sense of Humus.* Directed, filmed, and edited by Christopher Chapman. NFB Environment Program, 28 minutes (sound and colour), 1976.

NELSON, JAMIE. "Integrated Pest Management: Best Chance to Kick the Chemical Habit." *Not Man Apart* (Friends of the Earth), November-December, 1978, pp. 8-10.

"New Era in Weed Control." *Agri-book (Corn in Canada),* vol. 4, no. 3 (February, 1978), p. 67.

NORMINTON, CLAUDIA. *Pest Control in Canada: A Review.* Ottawa: Canadian Federation of Agriculture, May, 1973.

NORTH, F.K. "Oil and Gas: Surplus or Shortage?" Gerhard Herzberg Lecture (Carleton University, Ottawa), October 28, 1978.

* OELHAF, ROBERT C. *Organic Agriculture: Economic and Ecological Comparisons With Conventional Methods.* New York: John Wiley and Sons, 1978.

* Office of Technology Assessment (U.S. Congress). *Pest Management Strategies in Crop Protection.* Washington, D.C.: U.S. Government Printing Office, 1979.

"Organic Isn't Always Healthier." *The Globe and Mail,* October 19, 1978.

OLKOWSKI, WILLIAM "Editorial: The Magic Bullet vs. The IPM Approach." *The IPM Practitioner,* vol. 1, no. 8 (August, 1979), p. 2.

OLKOWSKI, WILLIAM, OLKOWSKI, HELGA, DARR, SHEILA, and REDMOND, JUDITH. "What is IPM?" *The IPM Practitioner,* vol. 1, no. 5 (May, 1979), p. 1.

RYDER, WALTER. "Integrated Pest Control and the Human Environment." *The Ecologist,* vol. 2, no. 3 (March, 1972), pp. 18-19.

SCHERESKY, ALVIN. "Problems Confronted When Making the Transition to Ecological Agriculture," In *Practical Alternatives to Chemicals in Agriculture.* Proceedings of a Conference held at Echo Valley Centre, Fort, Qu'Appelle, Saskatchewan, October 30-31, 1978. Regina, Sask.: University of Regina (Extension), 1979.

SCHRIEFER, DON. "Tillage Principles and Guidelines." *Acres U.S.A.,* December, 1979, p. 23.

* Science Council of Canada. *Canadian Food and Agriculture Sustainability and Self-reliance: A Discussion Paper.* Prepared by Science Council Committee for the Study: Canada's Scientific and Technological Contribution to World Food Supply. Ottawa: Science Council of Canada, May, 1979.

SHEA, KEVIN. "The Business of Biological Control." *Environment,* vol. 19, no. 3 (February, 1978), p. 67.

* SIEMENS, LEONARD B. "Ecological Agriculture: A Point of View." *Agrologist,* vol. 8, no. 4 (Fall 1979), pp. 12-15.

STONE, ALAN B. "Stuart Hill Calls It Ecological Agriculture; The Petro-Chemical Trust Calls It Preposterous." *Harrowsmith,* no. 2 (July-August, 1976), pp. 46-49, 60-61.

"Stored Grain Pests." *The IPM Practitioner,* vol. 2, no. 1 (January, 1980), p. 1.

* TUCKER, WILLIAM. "The Next American Dust Bowl. . .and How to Avert It." *Atlantic Monthly,* July, 1979, pp. 38-49.

TURNOCK, W.J. "Integrated Pest Control Programs." In *Chemicals and Agriculture: Problems and Alternatives.* Proceedings of a Seminar held at Echo Valley Centre, Fort Qu'Appelle, Saskatchewan, November 3-4, 1977. Regina, Sask.: Canadian Plains Research Center (University of Regina), 1978.

* United States Department of Agriculture (USDA Study Team on Organic Farming). *Report and Recommendations on Organic Farming.* Washington, D.C.: U.S. Government Printing Office, July, 1980.

VAIL, DAVID. "The Case for Organic Farming." *Science,* vol. 205 (13 July, 1979), pp. 180-181.

* VAN DEN BOSCH, ROBERT. *The Pesticide Conspiracy.* Garden City, N.Y.: Doubleday and Co. Inc., 1978.

VERITY-STEVENSON, BARBARA. "Fruit of the (Organic) Boom." *Harrowsmith,* no. 30 (1980), pp. 54-59, 88.

* WALTERS, CHARLES JR. and FENZAU, C.J. *An Acres U.S.A. Primer.* Raytown, Mo.: Acres, U.S.A., 1979.

WEBSTER, BAYARD. "Nice Bugs to the Rescue as Pesticides Lose Their Zip." *The New York Times,* September 23, 1979.

WOLF, RAY. "Winning the Bug War on the Toughest Battlefield." *Organic Gardening,* January, 1979, pp. 38-52.

ZWERDLING, DANIEL. "Curbing the Chemical Fix (Organic Farming: The Secret is it Works)." *The Progressive,* December, 1978, pp. 16-25.

———. *Organic Farming in California.* Canadian Broadcasting Corporation ("Radio Noon"), Toronto, January 3, 1980.

7 / The Barnyard Medicine Chest

Agriculture Canada. "DES Certification Requirements Lifted." News release, October 27, 1980.

———. *Mastitis Control.* Publication 1596. 1978.

"Antibiotics and Agriculture." *Agri-book (Herd Health and Hygiene),* September, 1979, p. 18.

* BOTTERELL, E.H. *Maintenance of Animal Health for Food Production* (Report of the Study of Animal Health Services in Ontario for the Ministry of Agriculture and Food). December, 1976.

* CAMPBELL, DAVID J. "Drug Residues in Animal Tissues and Their Regulatory Significance—The Canadian Point of View." *Journal of the Association of Official Analytical Chemists,* vol. 61, no. 5 (1978), pp. 1194-1197.

CARLYLE-GORDGE, PETER. "The Penicillin Hazard in the Everyday Hotdog." *Maclean's,* November 19, 1979, pp. 48f-48h.

* Committee on Antibiotics in Agriculture and the Protection of Human and Animal Health (W. D. Morrison, Chairman). *Report to the Minister of Agriculture & Food and Minister of Health (Ontario).* November, 1979.

CROSSLAND, JANICE. "Power to Resist." *Environment,* March, 1975, pp. 6-11.

"Firms Back Animal-Antibiotic Safety Tests." *Chemical Week,* April 30, 1980, p. 32.

* HAYS, V.W. and MUIR, W.M. "Efficacy and Safety of Feed Additive Use of Antibacterial Drugs in Animal Production." *Canadian Journal of Animal Science,* vol. 59 (June 1979), pp. 447-456.

* Health and Welfare Canada. "New Policy Concerning the Use of Antibiotics in Animal Feeds." News Release, July 7, 1977.

MARSHALL, ELIOT. "Antibiotics in the Barnyard." *Science,* April 25, 1980, pp. 376-379.

MORRISON, A.B. (Health Protection Branch, Health and Welfare Canada.) "Antibiotics in Feeds." Information Letter No. 505, October 14, 1977.

_____. "Antibiotics in Feeds." Information Letter No. 546. February 15, 1979.

_____. "Registered Medicated Feeds." Information Letter No. 561. August 9, 1979.

_____. "Sale of Veterinary New Drugs for Experimental Studies." Information Letter No. 578. January 31, 1980.

_____. "Withholding Times for Drugs Indicated for Use in Food-Producing Animals." Information Letter No. 402. November 14, 1973.

* MUNRO, I.C. and MORRISON, A.B. "Drug Residues in Foods of Animal Origin: Their Significance to Man." *Journal of the Association of Official Analytical Chemists,* vol. 53, no. 2 (1970), pp. 211-218.

* Office of Technology Assessment, Congress of the United States. *Drugs in Livestock Feeds, Volume I: Technical Report,* Washington, D.C.: Office of Technology Assessment, 1979.

OLESKIE, EDWARD. "Dairymen Must Control Antibiotic Residues." *Ontario Milk Producer,* June, 1977, pp. 17-19.

PAWLICK, THOMAS. "Arsenic and Old Layers." *Harrowsmith,* no. 32 (1980), pp. 21-22.

PIM, LINDA R. "One Man's Meat." *Probe Post,* September-October, 1979, p. 3.

REYNOLDS, NEIL. "Food Unfit for a Swine." *Harrowsmith,* no. 6 (1977), pp. 18-27, 80.

SCHOLLENBERG, E. and ALBRITTON, W.L. "Antibiotic Misuse in a Pediatric Teaching Hospital." *Canadian Medical Association Journal,* 12 January, 1980 (vol. 122), pp. 49-52.

* SHARBY, T.F. "Some Observations on the Use of Feed Additives in Canadian Animal Production." *Canadian Journal of Animal Science,* vol. 59 (June 1979), pp. 333-337.

SINGULAR, STEPHEN. "Saving People from Wonder Drugs." *Quest/80,* May, 1980, pp. 28-32, 90.

WALTON, SUSAN. "FDA Stymied on Antibiotic Ban." *BioScience,* May, 1980, pp. 295-296.

WELLFORD, HARRISON. "Behind the Meat Counter: The Fight Over DES." *Atlantic Monthly,* October, 1972, pp. 86-90.

ZWERDLING, DANIEL. "Drugs in the Meat Industry." *Ramparts,* June, 1973, pp. 37-41.

8 / Toxic Moulds in Natural Foods

* "Aflatoxin in Peanut Butter: A Tough Question." *Consumer Reports,* August, 1978, p. 437.

* Agriculture Canada. "Backgrounder: Fusarium Mold in Ontario and Quebec Wheat Crops." News Release, October 1, 1980.

* _____. *Ergot of Grains and Grasses*. Publication 1438, 1971.

Agriculture Canada/Canadian Grain Commission. *Official Grain Grading Guide, 1978 Edition*.

BRETT, ABIGAIL TRAFFORD. "Molds in the Corn." *The Washington Post*, June 15, 1975.

CAPORAEL, LINNDA R. "Ergotism: The Satan Loosed in Salem?" *Science*, April 2, 1976, pp. 21-26.

* CIEGLER, ALEX and BENNETT, J.W. "Mycotoxins and Mycotoxicoses." *BioScience*, August, 1980, pp. 512-515.

CLIFFORD, EDWARD. "Embargo Set on Wheat Sales While Tests Made for Toxin." *The Globe and Mail*, September 10, 1980.

_____. "First Commercial Crop of Peanuts Ends Decade-Long Research Effort." *The Globe and Mail*, February 23, 1980.

DECROSTA, ANTHONY. "Protect Yourself Against Aflatoxin." *Organic Gardening*, November, 1979, pp. 64-70.

FUNNELL, H.S. "Mycotoxins in Animal Feedstuffs in Ontario." *Canadian Journal of Comparative Medicine*, July, 1979, pp. 243-246.

* GOLDBLATT, LEO A. *Aflatoxin: Scientific Background, Control and Implications*. New York: Academic Press, 1969.

* Health and Welfare Canada. *Aflatoxins and Consumer Protection*. Dispatch No. 26. Revised 1979.

* LORENZ, KLAUS. "Ergot on Cereal Grains." *CRC Critical Reviews in Food Science and Nutrition*. May, 1979, pp. 311-354.

MARKS, H.L. and WYATT, R. D. "Genetic Resistance to Aflatoxin in Japanese Quail." *Science*, December 14, 1979, pp. 1329-1330.

"Mycotoxins—A Review." *Acres U.S.A.*, March, 1980, p. 16.

PIM, LINDA R. "Peanut Butter Woes." *Probe Post*, May-June, 1980, p. 3.

REYNOLDS, FRAN. "Baby-food Makers Told Not to Use Toxic Wheat." *The Toronto Star*, September 16, 1980.

* RODRICKS, JOSEPH V., HESSELTINE, CLIFFORD W., and MEHLMAN, MYRON A. *Mycotoxins in Human and Animal Health*. Park Forest South, Ill. Pathotox Publishers, Inc., 1977.

SCOTT, P.M. and KENNEDY, B.P.C. "Analysis and Survey of Ground Black, White and Capsicum Peppers for Aflatoxins." *Journal of the Association of Official Analytical Chemists*, November, 1973, pp. 1452-1457.

_____. "The Analysis of Spices and Herbs for Aflatoxins." *Canadian Institute of Food Science and Technology Journal*, vol. 8 no. 2 (1975), pp. 124-125.

STOLOFF, LEONARD and DALRYMPLE, BARBARA. "Aflatoxin and Zearalenone Occurrence in Dry-Milled Corn Products." *Journal of the Association of Official Analytical Chemists*, vol. 60, no. 3 (1977), pp. 579-582.

SPANOS, N.P. and GOTTLIEB, J. "Ergotism and the Salem Village Witch Trials." *Science,* December 24, 1976, pp. 1390-1394.

* TRENHOLM, H.L. and FARNWORTH, E.R. "Continuous Effort Prevents Growth of Moulds." *Canadian Consumer,* August, 1980, pp. 12-13.

WHITTAKER, JOHN. "Aflatoxins: A National Disgrace." *Countryside,* July, 1978, pp. 60-62.

YOUNG, J. CHRISTOPHER. "Ergot Contamination of Feedstuffs." *Feedstuffs,* 6 August, 1979, pp. 23-33.

9 / Organic Pollutants (Alphabet Soup)

Examples of Contamination

BRONSON, GAIL. "Study Finds Most Residents of Michigan Are Carrying Toxic Chemicals in Bodies." *Wall Street Journal,* October 9, 1978. (re: PBBs)

Canadian Broadcasting Corporation. "PBB Contamination." *Newsmagazine,* January 17, 1977.

"Contamination Found in Animal Feed, Food Across 17 States." *The Globe and Mail,* September 30, 1979. (re: PCBs)

____. "No Poison in Area Milk." *Ottawa Journal,* March 15, 1977. (re: PCP)

HEALD, HENRY. "No Poison in Area Milk." *Ottawa Journal,* March 15, 1977. (re: PCP)

Health and Welfare Canada/ Agriculture Canada. "PCB Investigation." News Release, October 2, 1979.

"Michigan Farmers Lose Bid for Compensation in PBB Contamination Case." *Not Man Apart* (Friends of the Earth), December-January 1978/79, p. 5.

PETERSON, IVER. "Michigan PBB: Not a Comedy, But Plenty of Errors." *The New York Times,* July 2, 1978.

PIM, LINDA R. "Chickens Fed PCB's: Bad Eggs Go to Market." *Probe Post,* September-October, 1979, p. 3.

"Everything Leaks"

* CASTRILLI, JOE and BLOCK, ELIZABETH. "Land Use: The Second Front in the War Over Great Lakes Water Quality." *Probe Post,* July-August, 1978, pp. 12-13.

* GLENN, BILL. "Toxic Winds are Poisoning Lakes." *Probe Post,* May-June, 1979, p. 2.

HENDERSON, DEBRA. "An Official Plea to Save Upper Lakes." *Probe Post,* July-August, 1979, p. 14.

MOCCIA, R.D., LEATHERLAND, J.F., and SONSTEGARD, R.A. "Increasing Frequency of Thyroid Gioters in Coho Salmon *(Oncorhynchus kisutch)* in the Great Lakes." *Science,* October 28, 1977, pp. 425-426.

MOCCIA, R.D., LEATHERLAND, J.F., SONSTEGARD, R.A. and HOLDRINET, M.V.H. "Are Goiter Frequencies in Great Lakes Salmon Correlated With Organochlorine Residues?" *Chemosphere,* vol. 8 (1978), pp. 649-652.

* PAEHLKE, ROBERT, "Guilty Until Proven Innocent: Carcinogens in the Environment." *Nature Canada,* April-June, 1980, pp. 18-23.

"Pollutants May Be Cause of Decline in Male Fertility." *The Globe and Mail,* September 12, 1979.

SONSTEGARD, R. and LEATHERLAND, J.F. "The Epizootiology and Pathogenesis of Thyroid Hyperplasia in Coho Salmon *(Oncorhynthus kisutch)* in Lake Ontario." *Cancer Research,* December, 1976, pp. 4467-4475.

PCBs in Food Other Than Fish

DAY, JAMES H. "PCB Disposal: A Medical Position." *Ontario Medical Review,* June, 1979, pp. 285-287.

* HIGHLAND, JOSEPH. "PCB's in Food." *Environment,* March, 1976, pp. 12-16.

KRUUS, P. and VALERIOTE, I.M. (editors). *Controversial Chemicals: A Citizen's Guide.* Montreal, Quebec: Multiscience Publications Ltd., 1979.

* MCLEOD, H.A., SMITH, D.C., and BLUMAN, N. "Pesticide Residues in the Total Diet in Canada, V: 1976 to 1978." *Journal of Food Safety,* 1980 (in press).

* National Research Council of Canada (Associate Committee on Scientific Criteria for Environmental Quality). *Polychlorinated Biphenyls: Biological Criteria for an Assessment of Their Effects on Environmental Quality.* NRCC Publication No. 16077. Ottawa: National Research Council of Canada, 1978.

* SASCHENBRECKER, P.W. "Levels of Terminal Pesticide Residues in Canadian Meat." *Canadian Veterinary Journal,* June, 1976, pp. 158-163.

SPEARS, JOHN. "Why Are PCBs So Dangerous?" *The Toronto Star,* November 4, 1979.

VERRETT, JACQUELINE and CARPER, JEAN. *Eating May Be Hazardous to Your Health.* New York: Simon & Schuster, 1974.

PCBs and DDT in Fish

"Anglers, Be Guided!" *Probe Post,* May-June, 1979, p. 5.

* FRANK, R. BRAUN, H.E., HOLDRINET, M., DODGE, D.P., and NEPSZY, S.J. "Residues of Organochlorine Insecticides and Polychlorinated Biphenyls in Fish from Lakes Saint Clair and Erie, Canada—1968-76." *Pesticides Monitoring Journal,* September, 1978, pp. 69-80.

* FRANK, R., HOLDRINET, M.V.H., BRAUN, H.E., DODGE, D.P., and SPRANGLER, G.E. "Residues of Organochlorine Insecticides and Polychlorinated Biphenyls in Fish from Lakes Huron and Superior, Canada—1968-76." *Pesticides Monitoring Journal,* September, 1978, pp. 60-68.

FRANK, R. HOLDRINET, M.V.H., DESJARDINE, R.L., and DODGE, D.P. "Organochlorine and Mercury Residues in Fish from Lake Simcoe, Ontario, 1970-76." *Environmental Biology of Fisheries,* vol. 3, no. 3 (1978). pp. 275-285.

* GIORNO, FRANK. "Fishing's Lure Dulled by Toxins." *Probe Post,* May-June, 1980, pp. 1-2.

"Let Them Eat Carp." *Probe Post,* July-August, 1978, p. 14.

* Ministry of the Environment (Ontario). *Health Implications of Contaminants in Fish.* 1978.

* Ministry of the Environment and Ministry of Natural Resources (Ontario). *Guide to Eating Ontario Sport Fish* (3 booklets: Great Lakes, Northern Ontario, Southern Ontario). Issued annually.

WORDSWORTH, ANNE. "It's a Pretty Kettle of Fish." *Probe Post,* July-August, 1979, p. 5. (re: "dumping" contaminated fish)

PCBs in Mother's Milk

Health and Welfare Canada. "Canadian Physicians Advised on Contaminated Breast Milk." News Release, May 24, 1978.

* MAKIN, KIRK. "PCB Guideline for Mother's Milk Called Too High." *The Globe and Mail,* May 7, 1980.

MALAREK, VICTOR. "Provincial Testing of Mother's Milk for PCB's Criticized by Biochemist." *The Globe and Mail,* May 31, 1979.

* MES, JOS and DAVIES, DAVID J. "Presence of Polychlorinated Biphenyl and Organochlorine Pesticide Residues and the Absence of Polychlorinated Terphenyls in Canadian Milk Samples." *Bulletin of Environmental Contamination and Toxicology,* vol. 21 (1979), pp. 381-387.

* MES, JOS, CAMPBELL, D.S., ROBINSON, R.N., and DAVIES, D.J.A. "Polychlorinated Biphenyl and Organochlorine Pesticide Residues in Adipose Tissue of Canadians." *Bulletin of Environmental Contamination and Toxicology,* vol. 17, no. 2 (1977), pp. 196-203.

"More Women Warned Eating Fish Has Risks." *The Globe and Mail,* July 21, 1978.

* MORRISON, A.B. "Polychlorinated Biphenyls, Department of National Health and Welfare—Committee Report." *Information Letter (Health Protection Branch),* March 31, 1978.

MYERS, A.W. "Every Baby Deserves the Breast." *Canadian Consumer,* February, 1979, pp. 12-14, 35.

* ROGAN, W.J., BAGNIEWSKA, A. and DAMSTRA, T. "Pollutants in Breast Milk." *The New England Journal of Medicine,* June 26, 1980, pp. 1450-1453.

United States Environmental Protection Agency. "PCB Chemicals in Human Breast Milk Reported by EPA." News Release, September 14, 1978.

"Wisdom of Breast-Feeding is Questioned After Contaminants Found in Mother's Milk." *The Globe and Mail,* September 29, 1977.

* YAKUSHIJI, T. WATANABE, I., KUWABARA, K., YOSHIDA, S., KOYAMA, K., HARA, I. and KUNITA, N. "Long-term Studies of the Excretion of Polychlorinated Biphenyls (PCBs) Through the Mother's Milk of an Occupationally Exposed Worker." *Archives of Environmental Contamination and Toxicology*, vol. 7 (1978), pp. 493-504.

* YAKUSHIJI, T. WATANABE, I., KUWABARA, K., YOSHIDA, S., KOYAMA, K., and KUNITA, N. "Levels of Polychlorinated Biphenyls (PCBs) and Organochlorine Pesticides in Human Milk and Blood Collected in Osaka Prefecture from 1972 to 1977." *International Archives of Occupational and Environmental Health*, vol. 43 (1979), pp. 1-15.

Coping with PCBs

CZERKAWSKI, E. and CONSTANTINE, T.A. "Alternative Transformers and Fluids Replace PCBs." *Electrical Business*, April, 1980, pp. 4, 8, 10.

"FDA Cuts PCBs Allowed in Food." *Environmental Defense Fund Letter*, July-August 1979.

GIBSON, BOB. "PCB Appeal Gets Tough." *Probe Post*, March-April, 1980, p. 2.

LANGFORD, H. DALE. "Looking at Polychlorinated Biphenyls as Environmental Object Lesson." *News Report* (National Academy of Sciences, National Academy of Engineering, Institute of Medicine and National Research Council (U.S.)), September, 1979, pp. 1, 4-5.

RUDOLPH, MARK S. "Road Oiling: An Example of Environmental Mismanagement." *Alternatives* (Friends of the Earth Canada), Spring, 1980, pp. 32-35.

Dioxins in Food

Agriculture Canada. "2,4-D Tests." News release, October 23, 1980.

BLATCHFORD, CHRISTIE. "Most Chicken Farmers Ignore Warning on Dioxin." *The Toronto Star*, December 9, 1980.

CARTER, C.D., KIMBROUGH, R.D., LIDDLE, J.A., CLINE, R.E., ZACK, M.M. JR., and BARTHEL, W.F. "Tetrachlorodibenzodioxin: An Accidental Poisoning Episode in Horse Arenas." *Science*, May 16, 1975, pp. 738-740.

* DICKSON, DAVID. "PCP Dioxins Found to Pose Health Risks." *Nature*, January 31, 1980, p. 418.

FERGUSON, JOCK and KEATING, MICHAEL. "Dioxin Levels in Lake Ontario Fish Among Highest in World." *The Globe and Mail*, December 10, 1980.

——. "Dioxin Traces Found in Chicken Livers." *The Globe and Mail*, December 9, 1980.

FOLSTER, DAVID. "Protest: The 2,4,5-T Herbicide Moves East and the Complaints Begin." *The Globe and Mail*, July 26, 1980.

* GALSTON, ARTHUR W. "Herbicides: A Mixed Blessing." BioScience, February, 1979, pp. 85-90.

GLENN, BILL. "Preserve Wood Without Preservatives." *Probe Post*, July-August, 1979, p. 16.

Health and Welfare Canada. "Dioxin Studies to Be Increased." News release, December 4, 1980.

MALAREK, VICTOR. "Testing for Dioxin is Late by 2½ Years." *The Globe and Mail*, April 28, 1979.

OHE, TAKESHI. "Pentachlorophenol Residues in Human Adipose Tissue." *Bulletin of Environmental Contamination and Toxicology*, vol. 22 (1979), pp. 287-292.

Pollution Probe. "Probe Says Dioxin Possible Threat to Ontario Drinking Water." News Release, April 26, 1979.

* "Polybrominated Biphenyls, Polychlorinated Biphenyls, Pentachlorphenol—and All That." (editorial). *The Lancet*, July 2, 1977, pp. 19-21.

POSNER, MICHAEL. "Agent Orange Comes Home to Do Battle." *Maclean's*, September 15, 1980. (re: teratogenicity of 2,4,5-T)

Public Broadcasting System. *A Plague on Our Children*. June 3, 1980. (re: 2,4,5-T, PCBs and dioxin)

RILEY, SUSAN. "Poison, Poison Everywhere." *Maclean's*, December 15, 1980.

SMITH, ELEANOR. "Agent Orange Veterans—Who is Responsible?" *Not Man Apart* (Friends of the Earth), September, 1980, pp. 13-14.

"Talk of the Town." *The New Yorker*, December 18. 1978. (re: TCDD in American food)

TOUBY, FRANK. "Dioxin's Deathly Taste Caught on the Fly." *Maclean's*, May 21, 1979, pp. 44b-44f.

* Vermont Public Interest Research Group. "Dioxins in Vermont." *Not Man Apart* (Friends of the Earth), August-September, 1978, pp. 6-7.

"What is Dioxin?" *Organic Gardening*, February, 1979, p. 126.

* ZABIK, MARY E., and ZABIK, MATTHEW J. "Dioxin Levels in Raw and Cooked Liver, Loin Steaks, Round, and Patties from Beef Fed Technical Grade Pentachlorophenol." *Bulletin of Environmental Contamination and Toxicology*, vol. 24 (1980), pp. 344-349.

Other Organics

CHEN, EDWIN. *PBB: An American Tragedy*. Inglewood Cliffs, New Jersey: Prentice Hall, Inc., 1979.

EGGINTON, JOYCE. *The Poisoning of Michigan*. New York: W.W. Norton, 1980.

* Environment Canada. "Polybrominated Biphenyls to be Banned." News Release, April 11, 1978.

Health and Welfare Canada. (Environmental Health Directorate, Health Protection Branch). *Benzene: Human Health Implications of Benzene at Levels Found in the Canadian Environment and Workplace*. Ottawa: Health and Welfare, Canada, March, 1979.

* KRUUS, P. and VALERIOTE, I.M. (editors). *Controversial Chemicals: A Citizen's Guide.* Montreal, Quebec: Multiscience Publications Ltd., 1979. (re: mirex, PBB's)

10 / Metals as Food Pollutants

Examples of Metal Contamination

National Academy of Sciences. *Marine Environmental Quality.* Washington, D.C.: National Academy of Sciences (Ocean Science Committee), 1971.

"No-Name Canned Pork Removed." *The Globe and Mail,* October 13, 1979.

"Sewage Sludge High in Chromium." *The Globe and Mail,* September 3, 1979.

TROYER, WARNER. *No Safe Place.* Toronto: Clarke, Irwin & Co. Ltd., 1977. (re: mercury pollution from Reed Paper Co.)

Metals, Metals Everywhere

* BOUDENE, CLAUDE. "Food Contamination by Metals." *In* Di Ferrante, Elvira (editor). *Trace Metals: Exposure and Health Effects.* Oxford, England: Pergamon Press, 1979. (Published for the Commission of the European Communities)

HUTCHINSON, T.C., CZUBA, M., and CUNNINGHAM, L. "Lead, Cadmium, Zinc, Copper and Nickel Distributions in Vegetables and Soils of an Intensely Cultivated Area and Levels of Copper, Lead and Zinc in the Growers." *In* Hemphill, D.D. (editor). *Trace Substances in Environmental Health— VIII.* Columbia, Missouri: University of Missouri, 1974.

LUCAS, H.F. and EDGINTON, D.N. "Concentrations of Trace Elements in Great Lakes Fishes." *Journal of the Fisheries Research Board of Canada,* vol. 27, no. 4 (1970), pp. 677-684.

OEHME, FREDERICK W. "Mechanisms of Heavy Metal Inorganic Toxicities." *In* Oehme, Frederick W. (editor). *Toxicity of Heavy Metals in the Environment.* New York: Marcel Dekker, Inc., 1978 (Part I) and 1979 (Part II).

* RUSSELL, LEON H., JR. "Heavy Metals in Foods of Animal Origin." *In* Oehme, Frederick W. (editor). *Toxicity of Heavy Metals in the Environment.* New York: Marcel Dekker, Inc., 1978 (Part I) and 1979 (Part II).

* SHACKLETTE, H.T., ERDMAN, J.A. HARMS, T.F., and PAPP, C.S.E. "Trace Elements in Plant Foodstuffs." *In* Oehme, Frederick W. (editor). *Toxicity of Heavy Metals in the Environment.* New York: Marcel Dekker, Inc., 1978 (Part I) and 1979 (Part II).

STOKES, PAMELA. "Heavy Metals—An Increasingly Serious Environmental Problem." *The Bulletin of the Conservation Council of Ontario,* October, 1977, pp. 13-15.

* UNDERWOOD, ERIC J. "Trace Elements." *In* National Academy of Sciences (Committee on Food Protection, Food and Nutrition Board, National Research Council). *Toxicants Occurring Naturally in Foods.* Washington, D.C.: National Academy of Sciences, 1973.

* ____. "Interactions of Trace Elements." *In* Oehme, Frederick W. (editor). *Toxicity of Heavy Metals in the Environment.* New York: Marcel Dekker, Inc., 1978 (Part I) and 1979 (Part II).

Sources and Toxicities of Metal Pollutants

FRANEY, PIERRE. "Cooking With Copper Risky With Some Acids." *The Globe and Mail,* January 16, 1980.

FRANK, R., BRAUN, H.E., ISHIDA, K., and SUDA, P. "Persistent Organic and Inorganic Pesticide Residues in Orchard Soils and Vineyards of Southern Ontario." *Canadian Journal of Soil Science,* November, 1976, pp. 463-484.

* FRANK, R., ISHIDA, K., and SUDA, P. "Metals in Agricultural Soils of Ontario." *Canadian Journal of Soil Science,* August, 1976, pp. 181-196.

FRANK, R., STONEFIELD, K.I. and SUDA, P. "Metals in Agricultural Soils of Ontario: II." *Canadian Journal of Soil Science,* May, 1979, pp. 99-103.

* KRUUS, P. and VALERIOTE, I.M. (editors). *Controversial Chemicals: A Citizen's Guide.* Montreal, Quebec: Multiscience Publications Ltd., 1979. (re: arsenic, cadmium, lead, mercury)

MACGREGOR, ALAN. "Analysis of Control Methods: Mercury and Cadmium Pollution." *Environmental Health Perspectives,* December, 1975, pp. 137-148.

MERANGER, J.C., CUNNINGHAM, H.M., and GIROUX, A. "Extraction of Heavy Metals from Plastic Food Containers: An X-Ray Fluorescence and Atomic Absorption Study." *Canadian Journal of Public Health,* July-August, 1974, pp. 292-296.

MUNRO, I.C. and MORRISON, A.B. "Drug Residues in Foods of Animal Origin: Their Significance to Man." *Journal of the Association of Official Analytical Chemists,* vol. 53, no. 2 (1970), pp. 211-218. (re: cadmium in livestock drugs)

* National Research Council of Canada (Associate Committee on Scientific Criteria for Environmental Quality). *Copper in the Aquatic Environment: Chemistry, Distribution and Toxicology.* NRCC Publication No. 16454. Ottawa: National Research Council of Canada, 1979.

* ____. *Effects of Arsenic in the Canadian Environment.* NRCC Publication No. 15391. Ottawa: National Research Council of Canada, 1978.

* ____. *Effects of Cadmium in the Canadian Environment.* NRCC Publication No. 16743. Ottawa: National Research Council of Canada, 1979.

* ____. *Effects of Chromium in the Canadian Environment.* NRCC Publication No. 15017. Ottawa: National Research Council of Canada, 1976.

*_____. *Effects of Lead in the Environment—1978: Quantitative Aspects.* NRCC Publication No. 16736. Ottawa: National Research Council of Canada, 1979.

*_____. *Effects of Mercury in the Canadian Environment.* NRCC Publication No. 16739. Ottawa: National Research Council of Canada, 1979.

ROSS. R.G. and STEWART, D.K.R. "Cadmium Residues in Apple Fruit and Foliage Following a Cover Spray of Cadmium Chloride." *Canadian Journal of Plant Science,* vol. 49 (1969), pp. 49-52.

Sources listed in preceding section.

STOEWSAND, G.S., STAMER, J.R., KOSIKOWSKI, F.V., MORSE, R.A., BACHE, C.A., and LISK, D.J. "Chromium and Nickel in Acidic Foods and By-Products Contacting Stainless Steel During Processing." *Bulletin of Environmental Contamination and Toxicology,* vol. 21 (1979), pp. 600-603.

TEMPLE, P.J., LINZON, S.N., and CHAI, B.L. "Contamination of Vegetation and Soil by Arsenic Emissions from Secondary Lead Smelters." *Environmental Pollution,* vol. 12 (1977), pp. 311-320.

WORDSWORTH, ANNE. ". . .And Not a Drop to Drink." *Canadian Consumer,* June, 1980, pp. 40-42. (re: drinking water as a pollution source for food)

Contamination by Circuitous Routes

CAMPBELL, MONI. "Centralized Sewage Systems Are a Drain on Capital." *Probe Post,* March-April, 1980, pp. 8-9.

* CHANEY, R.L. "Crop and Food Chain Effects of Toxic Elements in Sludges." In *Proceedings of the Joint Conference on Recycling Municipal Sludges and Effluents on Land.* Washington, D.C.: The National Association of State Universities and Land-Grant Colleges, 1974.

*"Fear of Sludging." *Harrowsmith,* no. 20 (1979), p. 99.

HANNAM, PETER. "The Energy Pinch and Food: Ideas Sprout on the Farms." *The Globe and Mail,* April 1, 1980.

* *Harrowsmith Staff Report.* "The Acid Earth." *Harrowsmith,* no. 27 (1980), pp. 32-41, 93.

HOWARD, ROSS and PERLEY, MICHAEL. *Acid Rain: The North American Forecast.* Toronto: House of Anansi, 1980.

Pollution Probe. *The Acid Rain Primer.* Toronto: Pollution Probe, 1980.

PASCA, T.M. "Confronting Air Pollution and Its Often Overlooked Impact on Crops and Forests." *Ceres* (FAO Report on Agriculture and Development), January-February, 1980, pp. 40-46.

SHEPPARD, ROBERT. "Acid Rain's Effects Spread to Trees, Crops." *The Globe and Mail,* October 16, 1979.

WELLER, PHIL and the Waterloo Public Interest Research Group. *Acid Rain: The Silent Crisis.* Waterloo, Ontario: Between the Lines, 1980.

Lead and Lead Again

* "Babies, Lead, and Evaporated Milk." *Consumer Reports,* May, 1980, p. 293.

* BENOY, C.J., HOOPER, P.A., and SCHNEIDER, R. "The Toxicity of Tin in Canned Fruit Juices and Solid Foods." *Food and Cosmetics Toxicology,* vol. 9 (1971), pp. 645-656.

BERITIC, T. and STAHULJAK, D. "Lead Poisoning from Lead-Glazed Pottery." *The Lancet,* March 25, 1961, p. 669.

COFFIN, D.E. and MCKINLEY, W.P. (Health and Welfare Canada). Unpublished data on chemical contaminants in the Canadian food supply, presented at the Fifth International Congress of Food Science and Technology, Kyoto, Japan, September, 1978.

FARROW, R.P., JOHNSON, J.H., GOULD, W.A., and CHARBONNEAU, J.E. "Detinning in Canned Tomatoes Caused by Accumulations of Nitrate in the Fruit." *Journal of Food Science,* vol. 36 (1971), p. 341.

HANKIN, L., HEICHEL, G.H. and BOTSFORD, R.A. "Lead Content of Printed Polyethylene Food Bags." *Bulletin of Environmental Contamination and Toxicology,* vol. 12, no. 6 (1974), pp. 645-648.

HAYES, WAYLAND, JR. *Toxicology of Pesticides.* Baltimore, Maryland: The Williams & Wilkins Co., 1975.

HOFF, J.E. and WILCOX, G.E. "Accumulation of Nitrate in Tomato Fruits and Its Effects on Detinning." *Journal of the American Society of Horticultural Science,* vol. 95 (1970), p. 92.

* KLEIN, M., NAMER, R., HARPUR, E., and CORBIN, R. "Earthenware Containers as a Source of Fatal Lead Poisoning: Case Study and Public Health Considerations." *The New England Journal of Medicine,* September 24, 1970, pp. 669-672.

* KOLBYE, A.C., MAHAFFEY, K.R., FIORINO, J.A., CORNELIUSSEN, P.C., and JELINEK, C.F. "Food Exposures to Lead." *Environmental Health Perspectives,* May, 1974, pp. 65-74.

* LEAH, T.D. *Environmental Contaminants Inventory Study No. 3: The Production, Use and Distribution of Lead in Canada.* (Report Series No. 41, Inland Waters Directorate). Ottawa: Environment Canada, 1976.

LINZON, S.N., CHAI, B.L., TEMPLE, P.J., PEARSON, R.G., and SMITH, M.L. "Lead Contamination of Urban Soils and Vegetation by Emissions from Secondary Lead Industries." *Journal of the Air Pollution Control Association,* July, 1976, pp. 650-654.

* MERANGER, J.C. "Lead in Ceramic Glazes." *Canadian Journal of Public Health,* September-October, 1973, pp. 472-476.

* MITCHELL, D.G. and ALDOUS, K.M. "Lead Content of Foodstuffs." *Environmental Health Perspectives,* May 1974, pp. 59-64.

* NEEDLEMAN, H.L., GUNNOE, C., LEVITON, A., REED, R., PERESIE, H., MAHER, C., and BARRETT, P. "Deficits in Psychologic and Classroom Performance of Children with Elevated Dentine Lead Levels." *The New England Journal of Medicine,* March 29, 1979, pp. 689-695.

* SETTLE, D.M. and PATTERSON, C.C. "Lead in Albacore: Guide to Lead Pollution in Americans." *Science,* March 14, 1980, pp. 1167-1176.

STRAUSS, MARINA. "Growing Lettuce Definitely Out, Neighbor to Lead Plant Decides." *The Globe and Mail,* June 12, 1979.

SULEK, A.M. "Evaluation of Lead in Raw and Canned Food." *Food Product Development,* July, 1978, pp. 61-62.

WAI, C.M., KNOWLES, C.R., and KEELY, J.F. "Lead Caps on Wine Bottles and Their Potential Problems." *Bulletin of Environmental Contamination and Toxicology,* vol. 21 (1979), pp. 4-6.

Mercury

"Ban on Lindane Contemplated by Authorities." *The Globe and Mail,* September 15, 1980.

* BRYAN, RORKE. *Much is Taken, Much Remains: Canadian Issues in Environmental Conservation.* North Scituate, Mass.: Duxbury Press, 1973.

FRANK, R., RAINFORTH, J.R., and SANGSTER, D. "Mushroom Production in Respect of Mercury Content." *Canadian Journal of Plant Science,* July, 1974, pp. 529-534.

JANKOWSKI, PAUL. "Fishery May Resume on Lake St. Clair." *The Globe and Mail,* January 1, 1980.

LOUGHTON, A. and FRANK, R. "Mercury in Mushrooms *(Agaricus bisporus)*." *In Mushroom Science IX (Part I)* (Proceedings of the Ninth International Scientific Congress on the Cultivation of Edible Fungi, Tokyo, 1974).

"Maximum Allowable Concentrations of Mercury Compounds." *Archives of Environmental Health,* December, 1969, pp. 891-905.

OZIEWICZ, STAN. "Costly Cleanup Called Only Remedy for Wabigoon." *The Globe and Mail,* July 10, 1980.

SAHA, J.G., LEE, Y.W., TINLINE, R.D., CHINN, S.H.F., and AUSTENSON, H.M. "Mercury Residues in Cereal Grains from Seeds or Soil Treated With Organomercury Compounds." *Canadian Journal of Plant Science,* September, 1970, pp. 597-599.

* SHERBIN, I. GRIFF (Environmental Protection Service, Environment Canada). *Mercury in the Canadian Environment.* (Economic & Technical Review Report EPS 3-EC-79-6) Ottawa: Environment Canada, April, 1979.

SMITH, GWEN. "Cancer-linked Pesticide Not Banned in Canada." *The Globe and Mail,* October 1, 1980. (re: lindane)

TEMPLE, P.J., and LINZON, S.N. "Contamination of Vegetation, Soil, Snow and Garden Crops by Atmospheric Deposition of Mercury from a Chlor-Alkali

Plant." *In* Hemphill, D.D. (editor). *Trace Substances in Environmental Health - XI: 1977.* Columbia, Mo.:University of Missouri, 1977.

Cadmium

* KIRKPATRICK, D.C. and COFFIN, D.E. "The Trace Metal Content of Representative Canadian Diets in 1970 and 1971." *Canadian Institute of Food Science and Technology Journal,* vol. 7, no. 1 (1974), pp. 56-58.
* _____. "The Trace Metal Content of a Representative Canadian Diet in 1972." *Canadian Journal of Public Health,* March-April, 1977, pp. 162-164.
* MERANGER, J.C. and SMITH, D.C. "The Heavy Metal Content of a Typical Canadian Diet." *Canadian Journal of Public Health,* January-February, 1972, pp. 53-57.

Total Metal Intake and "Safe" Levels

* MAHAFFEY, K.R. CORNELIUSSEN, P.E., JELINEK, C.F., and FIORINO, J.A. "Heavy Metal Exposure from Foods." *Environmental Health Perspectives,* December, 1975, pp. 63-69.
Sources listed in preceding section.

Canadian Regulations on Metals in Food

* *The Food and Drugs Act and Regulations.* Supply and Services Canada.
* *The Hazardous Products Act and Regulations.* Supply and Services Canada.
* RADER, W.A. and SPAULDING, J.E. "Regulatory Aspects of Trace Elements in the Environment." *In* Oehme, Frederick W. (editor). *Toxicity of Heavy Metals in the Environment.* New York: Marcel Dekker, Inc., 1978 (Part I) and 1979 (Part II).

11 / A Potpourri

Plastic Packages

"Benzene Under the CPSC [Consumer Product Safety Commission] Gun." *Modern Packaging,* March, 1978, p. 4.

COFFIN, D.E. and MCKINLEY, W.P. (Health and Welfare Canada). Unpublished data on chemical contaminants in the Canadian food supply, presented at the Fifth International Congress of Food Science and Technology, Kyoto, Japan, September, 1978.

COKERBY, E.T. "Thermoplastic Polyester May Be Key to Search for Safe Family-Size Bottle." *The Globe and Mail,* November 12, 1979.

FLAVIN, CHRISTOPHER. "Plastic Becoming More Essential." *The Toronto Star,* July 10, 1980.

HARE, MICHAEL J. *Carbonated Soft Drink Packaging in Ontario: An Environmental Reappraisal.* Toronto: Waste Management Advisory Board (Government of Ontario), August, 1979.

* KRUUS, P. and VALERIOTE, I.M. (editors) *Controversial Chemicals: A Citizen's Guide.* Montreal, Quebec: Multiscience Publications Ltd., 1979. (re: vinyl chloride and polyvinyl chloride)

* LEAH, T.D. *Environmental Contaminants Inventory Study No. 4: The Production, Use and Distribution of Phthalate Acid Esters in Canada.* (Report Series No. 47). Burlington, Ontario: Inland Waters Directorate (Ontario Region), Fisheries and Environment Canada, 1977.

* LEFAUX, RENE. *Practical Toxivology of Plastics.* (Translated by Scripta Technica Ltd.; English edition edited by Peter P. Hopf). London: Iliffe Books Ltd., 1968.

"Lemon Tea, Foam-like Cups Don't Mix: MD." *The Globe and Mail,* November 5, 1979.

* MORRISON, A.B. (Health Protection Branch, Health and Welfare Canada). "Food Packaging Materials Containing Acrylonitrile Residues." Information Letter No. 586, September 19, 1980.

* PHILLIPS, MICHAEL. "Lemon-Tea Drinkers—A Group at Risk?" *The New England Journal of Medicine,* November 1, 1979, pp. 1005-1006.

PIM, LINDA R. "Packaged Peril." *Probe Post,* May-June, 1979, p. 4.

PINTAURO, NICHOLAS D. *Food Packaging.* Park Ridge, N.J.: Noyes Data Corporation, 1978.

PINTO, ALLEN. "The Federal Ban Wagon Gains Momentum." *Modern Packaging,* June, 1978, pp. 21-25. (re: benzene, PVC, acrylonitrile)

"Second Attempt Made to Ban Plastic Bottles." *The Globe and Mail,* September 22, 1977.

SQUIRELL, D.C.M. and LYONS, W.T. "Methods for the Measurement of Vinyl Choride in Polyvinyl Chloride, Air, Water and Foodstuffs". Volume 2 of: Egan, H. (editor). *Environmental Carcinogens—Selected Methods of Analysis.* International Agency for Research on Cancer (Scientific Publications No. 22), 1978.

PCBs in Food Packages

CORNELIUSSEN, PAUL E. (Division of Chemical Technology, Bureau of Foods, Food and Drug Administration). *Testimony on PCB's in Paper Food Packaging Materials.* Washington, D.C.: Food and Drug Administration, July 1, 1975. (Docket No. 75-N-0013)

* VILLENEUVE, D.C., REYNOLDS, L.M., THOMAS, G.H., and PHILLIPS, W.E.J. "Polychlorinated Biphenyls and Polychlorinated Terphenyls in Canadian Food Packaging Materials." *Journal of the Association of Official Analytical Chemists,* July, 1973, pp. 999-1001.

WESSEL, JOHN R. (Scientific Coordinator, Office of the Associate Commissioner for Compliance, Food and Drug Administration). *Testimony on PCB's in Paper Food Packaging Materials.* Washington, D.C.: Food and Drug Administration, October 16, 1975. (Docket No. 75-N-0013)

Asbestos

"Asbestos Fibres Found in Tests on 15 Wines." *The Globe and Mail,* June 14, 1977.

* CUNNINGHAM, H.M. and PONTEFRACT, R.D. "Asbestos Fibres in Beverages and Drinking Water." *Nature,* vol. 232 (1971), pp. 332-333.

* _____. "Asbestos Fibres in Beverages, Drinking Water and Tissues: Their Passage Through the Intestinal Wall and Movement Through the Body." *Journal of the Association of Official Analytical Chemists,* vol. 56, no. 4 (1973), pp. 976-981.

_____. "Penetration of Asbestos Through the Digestive Tract of Rats." *Nature,* vol. 243 (1973), pp. 352-353.

CUNNINGHAM, H.M., MOODIE, C.A., LAWRENCE, G.A., and PONTEFRACT, R.D. "Chronic Effects of Ingested Asbestos in Rats." *Archives of Environmental Contamination and Toxicology,* vol. 6 (1977), pp. 507-513.

KRUSS, P. and VALERIOTE, I.M. (editors) *Controversial Chemicals: A Citizen's Guide.* Montreal, Quebec: Multiscience Publications Ltd., 1979.

* National Research Council of Canada (Associate Committee on Scientific Criteria for Environmental Quality). *Effects of Asbestos in the Canadian Environment.* (NRCC Publication No. 16452.) Ottawa: National Research Council of Canada, 1979.

* "Test: Asbestos in Wine." *Canadian Consumer,* June, 1977, pp. 44-46.

Radiation

ABERNETHY, SCOTT. "Chips that Glow in the Night." *Harrowsmith,* no. 29 (1980), pp. 11-12.

* BAKER, B.E., LAUER, B.H. and SAMUELS, E.R. "Strontium-90 and Cesium-137 Levels in the Milks of Some Arctic Species." *Journal of Dairy Science,* vol. 51, no. 9 (1968), pp. 1508-1510.

BROSS, I.D.J. and NATARAJAN, N. "Leukemia from Low-Level Radiation." *The New England Journal of Medicine,* July 20, 1972, pp. 107-110.

* CALDICOTT, HELEN M. *Nuclear Madness: What You Can Do!* Brookline, Mass.: Autumn Press, Inc., 1978.

Department of Commerce (U.S.). *Current Status and Commercial Prospects for Radiation Preservation of Food.* (Prepared for Division of Isotopes Development, U.S. Atomic Energy Commission.) Washington, D.C.: U.S. Government Printing Office, 1965.

* Environmental Defense Fund and Robert H. Boyle. *Malignant Neglect*. New York: Alfred A. Knopf, 1979.

"Irradiated Food: Industry Warms Up to an Old Idea." *Chemical Week*, October 8, 1980, pp. 42-43.

MACAULAY, HUGH. (Chairman, Ontario Hydro.) "No High Tomatoes." *The Globe and Mail*, June 9, 1980.

MCLAREN, CHRISTIE. "Radiation Rate is High in Tomatoes Produced in Hydro 'A-Plant' Study." *The Globe and Mail*, May 22, 1980.

PAWLICK, THOMAS. "Tuber Syndrome?" *Harrowsmith*, no. 25 (1980), pp. 19-21.

* _____. "The Silent Toll: Uncovering the Deadly Consequences of Three Mile Island." *Harrowsmith*, no. 28 (1980), pp. 33-49.

RUBIN, NORMAN. "Carbon-14 Under Debate." *Probe Post*, November-December, 1979, p. 2.

_____. "Radioactive Carbon-14 May Get Major Study." *Probe Post*, July-August, 1980, p. 5.

SWANN, MARK. "The Hazards of Ionizing Radiation." *Rodale's Environmental Action Bulletin*, October 30, 1976, pp. 1-8.

"Tritium in DNA Threat." *Probe Post*, July-August, 1979, p. 10.

URROWS, GRACE M. *Food Preservation by Irradiation*. Oak Ridge, Tenn.: United States Atomic Energy Commission, Division of Technical Information Extension, 1964.

Appendix A / Toxins on the Job: Occupational Exposure to Pesticides

Agriculture Canada. *Pesticides: Their Implications for Agriculture*. Publication 1518. 1973.

Agriculture Canada, Fisheries and Environment Canada, and Health and Welfare Canada. *Pesticide Use and Control in Canada*. Prepared for: The Canadian Council of Resource and Environment Ministers Meeting of June 1-2, 1977. Ottawa: Minister of Supply and Services, February, 1978.

ASHFORD, NICHOLAS A. *Crisis in the Workplace*. Cambridge, Mass.: The MIT Press, 1979 (revised).

BATES, J.A.R. "Hazards from Inadequate/Improper Labelling and Packaging of Pesticides." *FAO Plant Protection Bulletin*, vol. 26 (1978), pp. 123-128.

* BRANDIS, W.B.A. *Pesticides and Their Safe Use*. Ottawa: Canadian Agricultural Chemicals Association, 1975.

* BROWN, J.R., CHAI, F.C., CHOW, L.Y., and STOPPS, G.J. "Human Blood Cholinesterase Activity—Holland Marsh, Ontario, 1976." *Bulletin of Environmental Contamination and Toxicology*, vol. 19 (1978), pp. 617-623.

CARMAN, G.E. "Worker Reentry Safety, I: An Overview of the Reentry Problem on Citrus in California." *Residue Reviews*, vol. 62 (1976), pp. 1-6.

* CULVER, B. DWIGHT. "Worker Reentry Safety, VI: Occupational Health Aspects of Exposure to Pesticide Residues." *Residue Reviews,* vol. 62 (1976), pp. 41-44.

GUNTHER, F.A., IWATA, Y., CARMAN, G.E., and SMITH, C.A. "The Citrus Reentry Problem: Research on its Causes and Effects, and Approaches to its Minimization." *Residue Reviews,* vol. 67 (1977), pp. 1-132.

* HAYES, WAYLAND, JR. *Toxicology of Pesticides.* Baltimore, Md.: The Williams Wilkins Co., 1975.

KAHN, EPHRAIM. "Outline Guide for Performance of Field Studies to Establish Safe Reentry Intervals for Organophosphate Pesticides." *Residue Reviews,* vol. 70 (1979), pp. 27-43.

KLEMMER, HOWARD W. "Human Health and Pesticides—Community Pesticide Studies." *Residue Reviews,* vol. 41 (1972), pp. 55-63.

KRAYBILL, H.F. "Significance of Pesticide Residues in Foods in Relationship to Total Environmental Stress." *Canadian Medical Association Journal,* January 25, 1969, pp. 204-215.

MACKENZIE, C.J.G., OLDHAM, W.K., and POWRIE, W.D. *Royal Commission of Inquiry Into the Use of Pesticides and Herbicides, Final Report of the Commissioners, May 30, 1975* (British Columbia). Volumes 1, 2 and 3. 1975.

MAZZORA, MARIA. "Farmworkers Live Under the Spray Gun." *Not Man Apart* (Friends of the Earth), September, 1980, p. 17.

* MORGAN, D.P., LIN, L.I., and SAIKALY, H.H. "Morbidity and Mortality in Workers Occupationally Exposed to Pesticides." *Archives of Environmental Contamination and Toxicology,* vol. 9 (1980), pp. 349-382.

* National Research Council of Canada (Associate Committee on Scientific Criteria for Environmental Quality). *Carbofuran: Criteria for Interpreting the Effects of Its Use on Environmental Quality.* NRCC Publication no. 16740. Ottawa: National Research Council of Canada, 1979.

* ____. *Phenoxy Herbicides: Their Effects on Environmental Quality (With Accompanying Scientific Criteria for 2,3,7,8-Tetrachlorodibenzo-p-Dioxin [TCDD]).* NRCC Publication no. 16075. Ottawa: National Research Council of Canada, 1978.

NORMINTON, CLAUDIA. *Pest Control in Canada: A Review.* Ottawa: Canadian Federation of Agriculture, May, 1973.

* Ontario Ministry of the Environment. *Pesticides Safety Handbook.* October 1979.

* *The Pest Control Products Act and Regulations.* Supply and Services Canada.

POPENDORF, W.J., SPEAR, R.C., LEFFINGWELL, J.T., YAGER, J., and KAHN, E. "Harvester Exposure to Zolone (Phosalone) Residues in Peach Orchards." *Journal of Occupational Medicine,* vol. 21, no. 3 (March 1979), pp. 189-194.

SHAFIK, M.T. and BRADWAY, D.E. "Worker Reentry Safety, VIII: The Determination of Urinary Metabolites—An Index of Human and Animal Exposure to Nonpersistent Pesticides." *Residue Reviews,* vol. 62 (1976), pp. 59-77.

* SHEA, KEVIN. "Profile of a Deadly Pesticide." *Environment,* January-February, 1977, pp. 6-12. (re: leptophos)

SMYTHE, JENNIE (Saskatchewan Department of Labour). "Health Effects of Agricultural Chemicals." *In Chemicals and Agriculture: Problems and Alternatives.* Proceedings of a Seminar Held at Echo Valley Centre, Fort Qu'Appelle, Saskatchewan, November 3-4, 1977. Regina, Sask.: Canadian Plains Research Center (University of Regina), 1978.

SPEAR, R.C., POPENDORF, W.J., LEFFINGWELL, J.T., MILBY, T.H. DAVIES, J.E., and SPENCER, W.F. "Fieldworkers' Response to Weathered Residues of Parathion." *Journal of Occupational Medicine,* vol. 19, no. 6 (June 1977), pp. 406-410.

SPEAR, R.C., POPENDORF, W.J., SPENCER, W.F., and MILBY, T.H. "Worker Poisoning Due to Paraoxon Residues." *Journal of Occupational Medicine,* vol. 19, no. 6 (June 1977), pp. 411-414.

WARE, GEORGE W. and MORGAN, DONALD P. "Worker Reentry Safety, IX: Techniques of Determining Safe Reentry Intervals for Organophosphate-Treated Cotton Fields." *Residue Reviews,* vol. 62 (1976), pp. 79-100.

WEIR, DAVID, SCHAPIRO, MARK, and JACOBS, TERRY. "The Boomerang Crime." *Mother Jones,* November, 1979, pp. 40-48.

WHORTON, D., MILBY, T.H., KRAUSS, R.M., and STUBBS, H.A. "Testicular Function in DBCP Exposed Pesticide Workers." *Journal of Occupational Medicine,* vol. 21, no. 3 (March 1979), pp. 161-166.

ZWERDLING, DANIEL. "The New Pesticide Threat." *In* Lerza, Catherine and Jacobson, Michael (editors). *Food for People, Not for Profit.* New York: Ballantine Books, 1975.

APPENDIX D
For Further Information

Books

BRYAN, RORKE. *Much is Taken, Much Remains: Canadian Issues in Environmental Conservation.* North Scituate, Mass.: Duxbury Press, 1973.

DICKEY, LAWRENCE D. (editor). *Clinical Ecology.* Springfield, Ill.: Charles C. Thomas, Publisher, 1976.

Environmental Defense Fund and Robert H. Boyle. *Malignant Neglect.* New York: Alfred A. Knopf, 1979.

ESTRIN, DAVID and SWAIGEN, JOHN. *Environment on Trial: A Handbook of Ontario Environmental Law* (revised and expanded). Revised edition edited by Mary Anne Carswell and John Swaigen. Toronto: Macmillan of Canada Ltd., 1978.

EPSTEIN, SAMUEL S. *The Politics of Cancer* (revised and expanded edition). Garden City, N.Y.: Anchor Press/Doubleday, 1979.

GALLI, C.L., PAOLETTI, P., and VETTORAZZI, G. (editors). *Chemical Toxicology of Food.* Amsterdam: Elsevier/North-Holland Biomedical Press, 1978.

HALL, ROSS HUME. *Food for Nought: The Decline in Nutrition.* New York: Harper and Row, 1974.

Harrowsmith (editors). *The Canadian Whole Food Book.* Camden East, Ont.: Camden House Publishing, 1980.

HUNTER, BEATRICE TRUM. *Consumer Beware! Your Food and What's Been Done to It.* New York: Bantam Books, 1971.

_____. *The Mirage of Safety: Food Additives and Federal Policy.* New York: Charles Scribner's Sons, 1975.

LAWRENCE, JAMES (editor). *The Harrowsmith Sourcebook.* Camden East, Ont.: Camden House Publishing, 1979.

MACKARNESS, RICHARD. *Chemical Victims.* London: Pan Books. 1980.

National Research Council of Canada (Associate Committee on Scientific Criteria for Environmental Quality). A series of publications dealing with individual environmental contaminants: Fluoride (NRCC Pub. No. 12226, 1971, and No. 16081, 1978); Lead (No. 13682, 1973 and No. 16736, 1979); Chlordane (No. 14094, 1974); Endosulfan (No. 14098, 1975); Methoxychlor (No. 14102, 1975); Chromium (No. 15017, 1976); *Bacillus thuringiensis* (No. 15385, 1976); Arsenic (No. 15391, 1978); Phenoxy Herbicides (No. 16075, 1978); Polychlorinated Biphenyls (No. 16077, 1978); Chlorpyrifos (No. 16079, 1978); Asbestos (No. 16452, 1979); Copper (No. 16454, 1979); Mercury (No. 16739, 1979); Carbofuran (No. 16740, 1979); Cadmium (No. 16743,1979).

PIM, LINDA R. *Additive Alert: A Guide to Food Additives for the Canadian Consumer.* Toronto: Doubleday Canada, 1979.

RANDOLPH, THERON. *Human Ecology and Susceptibility to the Chemical Environment.* Springfield, Ill.: Charles C. Thomas, Publisher, 1962.

SMALL, BRUCE and SMALL, BARBARA. *Sunnyhill: The Health Story of the 80's.* Goodwood, Ont.: Small and Associates, Publishers, 1980.

Periodicals

(A) Popular

Acres U.S.A. (A Voice for Eco-Agriculture). 10008 East 60th. Terrace, Raytown, Missouri, U.S.A., 64133.

Canadian Consumer. 200 First Avenue, Ottawa, Ontario, K1S 2G6.

Consumer Reports. 256 Washington Street, Mount Vernon, New York, U.S.A., 10550.

Environment. 4000 Albemarle Street N.W., Washington, D.C. U.S.A., 20016.

Harrowsmith. Camden East, Ontario, K0K 1J0.

The IPM Practitioner (The Newsletter of Integrated Pest Management). Bio-Integral Resource Center, 1805 2nd. Street, Berkeley, California, U.S.A., 94706.

The New Farm. Rodale Press, Inc., 33 East Minor Street, Emmaus, Pennsylvania, U.S.A., 18049.

Not Man Apart. 124 Spear Street, San Francisco, California, U.S.A., 94105.

Organic Gardening. Rodale Press, Inc., 33 East Minor Street, Emmaus, Pennsylvania, U.S.A., 18049.

Probe Post. 12 Madison Avenue, Toronto, Ontario, M5R 2S1.

(B) Scientific

Archives of Environmental Contamination and Toxicology. Springer-Verlag, 175 Fifth Avenue, New York, New York, U.S.A., 10010.

Archives of Environmental Health. 4000 Albemarle Street N.W., Washington, D.C., U.S.A., 20016.

BioScience. 1401 Wilson Blvd., Arlington, Virginia, U.S.A., 22209.

Bulletin of Environmental Contamination and Toxicology. Springer-Verlag, 175 Fifth Avenue, New York, New York, U.S.A., 10010.

Canadian Institute of Food Science and Technology Journal. 46 Elgin Street, #38, Ottawa, Ontario, K1P 5K6.

Canadian Journal of Public Health. 1335 Carling Avenue, #210, Ottawa, Ontario, K1Z 8N8.

Canadian Medical Association Journal. P.O. Box 8650, Ottawa, Ontario, K1G 0G8.

Environmental Health Perspectives. National Institute of Environmental Health Sciences, Box 12233, Research Triangle Park, North Carolina, U.S.A., 27709.

Food and Cosmetics Toxicology. Pergamon Press, Headington Hill Hall, Oxford 0X3 0BW, England.

Food Drug Cosmetic Law Journal. 4025 W. Peterson Avenue, Chicago, Illinois, U.S.A., 60646.

International Archives of Occupational and Environmental Health. Springer-Verlag, 175 Fifth Avenue, New York, New York, U.S.A., 10010.

Journal of the Association of Official Analytical Chemists. Box 540, Benjamin Franklin Station, Washington, D.C., U.S.A., 20044.

Journal of Food Protection. 413 Kellogg Avenue, P.O. Box 701, Ames, Iowa, U.S.A., 50010.

The Lancet. 7 Adams Street, Adelphi, London WC2N 6AD, England.

Nature. 4 Little Essex Street, London WC2R 3LF, England.

New England Journal of Medicine. 10 Shattuck, Boston, Massachusetts, U.S.A., 02115.

New Scientist. King's Reach Tower, Stamford Street, London SE1 9LS, England.

Pesticides Monitoring Journal. Technical Services Division, Office of Pesticide Programs, Environmental Protection Agency, Washington, D.C., U.S.A., 20460.

Pesticide Science. Society of Chemical Industry, 14 Belgrave Square, London SW1X 8PS, England.

Residue Reviews. Springer-Verlag, 175 Fifth Avenue, New York, New York, U.S.A., 10010.

Science. 1515 Massachusetts Avenue N.W., Washington, D.C., U.S.A., 20005.

Organizations Concerned With Environment and Health

Canadian Environmental Law Association, 8 York Street, 5th. Floor S., Toronto, Ontario, M5J 1R2.

Canadian Organic Certification Association, c/o Greenleaf Whole Foods Ltd., 117 Weber Street West, Kitchener, Ontario N2H 3Z8.

Canadian Organic Growers, 33 Karnwood Drive, Scarborough, Ontario, M1L 2Z4.

Canadian Public Health Association, 1335 Carling Avenue, #210, Ottawa, Ontario, K1Z 8N8. (Publishers of the *Canadian Journal of Public Health* and Canadian distributor for publications of the World Health Organization)

Center for Science in the Public Interest, 1755 S Street N.W., Washington, D.C., U.S.A., 20009.

Consumers' Association of Canada, 251 Laurier Avenue W., #801, Ottawa, Ontario, K1P 5Z7. (Publishers of *Canadian Consumer*)

Consumers Union, 256 Washington Street, Mount Vernon, New York, U.S.A., 10550. (Publishers of *Consumer Reports*)

Ecological Agriculture Project, Macdonald College (McGill University), Ste Anne De Bellevue, Quebec, H9X 1C0.

Environmental Defense Fund, 1525 18th. Street N.W., Washington, D.C., U.S.A., 20036.

Friends of the Earth, 124 Spear Street, San Francisco, California, U.S.A., 94105. (Publishers of *Not Man Apart*)

Human Ecology Foundation of Canada, R.R. #1, Goodwood, Ontario, L0C 1A0. (re: obtaining chemical-free meats)

International Federation of Organic Agriculture Movements, c/o E. Coleman, 9 Hill Street, Topsfield, Massachusetts, U.S.A., 01983.

Joint Project for Sustainable Agriculture, 1747 Connecticut Avenue N.W., Washington, D.C., U.S.A., 20009.

Ontario Public Interest Research Group (Provincial Office), 121A Avenue Road, Toronto, Ontario, M5R 2G3.

Pollution Probe, University of Toronto, Toronto, Ontario, M5S 1A1. (Publishers of the *Probe Post*)

West Coast Environmental Law Association, 207 West Hastings, #1012, Vancouver, British Columbia, V6B 1J9.

Canadian Agriculture and Food Industry Associations

Agricultural Institute of Canada, 151 Slater Street, #907, Ottawa, Ontario, K1P 5H4. (Publishers of *The Agrologist*)

Canadian Agricultural Chemicals Association, 116 Albert Street, #710, Ottawa, Ontario K1P 5G3.

Canadian Federation of Agriculture, 111 Sparks Street, Ottawa, Ontario, K1P 5B5.

Canadian Food Processors Association, 130 Albert Street, #1409, Ottawa, Ontario, K1P 5G4.

Canadian Institute of Food Science and Technology, 46 Elgin Street, #38, Ottawa, Ontario, K1P 5K6.

Grocery Products Manufacturers of Canada, 797 Don Mills Road, Don Mills, Ontario, M3C 1V1.

National Farmers Union, 250C 2nd. Avenue South, Saskatoon, Saskatchewan, S7K 2M1.

Government Agencies

Canada (Federal)

Consumer Standards Directorate, Consumer and Corporate Affairs Canada, Ottawa, Ontario, K1A 0C9.

Environmental Protection Service, Environment Canada, Ottawa, Ontario, K1A 1C8.

Health of Animals Branch, Agriculture Canada, Ottawa, Ontario, K1A 0C5.

Health Protection Branch, Health and Welfare Canada, Ottawa, Ontario, K1A 0L2.

Plant Products Division, Food Production and Marketing Branch, Agriculture Canada, Ottawa, Ontario, K1A 0C5.

Provincial

Each province of Canada has its own departments of agriculture, environment, health and consumer affairs. Consult your local telephone directory.

U.S.A. (Federal)

Department of Agriculture (USDA), Independence Avenue between 12th. and 14th. Streets S.W., Washington, D.C., U.S.A., 20250.

Environmental Protection Agency (EPA), 401 M Street S.W., Washington, D.C., U.S.A., 20024.

Food and Drug Administration (FDA), 5600 Fishers Lane, Rockville, Maryland, , U.S.A., 20857.

Miscellaneous

Canadian Broadcasting Corporation (CBC). "Radio Noon." 12 noon to 2 p.m., Monday to Friday, local time; locally produced.

Canadian Broadcasting Corporation (CBC). "The Food Show." 8:30 to 9:00 a.m., Sunday, local time; produced in Toronto.

Index